"Is This Thing On?"

Sound Systems for Your
Business, School, and Auditorium

By
Gordon McComb

PROMPT.
PUBLICATIONS

An Imprint of
Howard W. Sams & Company
Indianapolis, Indiana

REVISED FIRST EDITION, 1996

PROMPT® Publications is an imprint of Howard W. Sams & Company, a Bell Atlantic Company, 2647 Waterfront Parkway, E. Dr., Suite 300, Indianapolis, IN 46214-2041.

This book was originally developed and published as *Installing and Maintaining Sound Systems* by Master Publishing, Inc., 522 Cap Rock Drive, Richardson, Texas 75080-2036.

International Standard Book Number: 0-7906-1081-7

Editors: Charles Battle, Gerald Luecke
Text Design and Artwork: Plunk Design, Dallas, TX
Cover Design by: Phil Velikan

Trademark Acknowledgments:
- Dolby® and Prologic® are registered trademarks of Dolby Laboratories Licensing Corporation.
- PEM™ is a trademark of Crown International.
- Radio Shack® is a registered trademark of Tandy Corporation.
- RCA® is a registered trademark of Radio Corporation of America.

All terms in this book that are known or suspected to be trademarks or services have been appropriately capitalized. PROMPT® Publications, Howard W. Sams & Company, and Bell Atlantic cannot attest to the accuracy of this information. Use of a term in this book should not be regarded as affecting the validity of any trademark or service mark.

Printed in the United States of America

9 8 7 6 5 4 3 2 1

Table of Contents

Preface

"Is This Thing On?" has been written principally for all of you who have responsibility for sound systems in churches, schools, city offices, hotels, and auditoriums. Whether you are a custodian, building engineer, technician or "do-it-yourselfer," there is an indicated need for information on installing and maintaining a very significant and important system—the sound system. Whether you have responsibility for designing, purchasing, installing, maintaining, or operating sound systems, there are answers to your questions in this book. You will find answers on room acoustics, the best size for the speaker systems and power amplifier for a particular room; the type, use and placement of microphones and speakers; how components of a sound system are specified; the type connectors that are used and how sound systems are installed and connected together electrically. Only sound systems that are installed in rooms holding up to a maximum of 1200 people are discussed. Systems for larger rooms and outdoor installations require special equipment and engineering knowledge that are beyond the scope of this book.

"Is This Thing On?" begins by explaining different types of sound systems, basic components, and general information on tools and safety. It then deals with the basics of acoustics, discusses how sound is affected by an enclosed room, explains acoustic feedback and reviews common sound system terminology—which includes understanding the different decibel measurement references.

There is a chapter on component specifications to help in designing and purchasing the proper parts for your systems, followed by chapters on installing, wiring, testing and adjusting sound systems. Frequency response, power ratings, impedance matching, wire size for interconnections, routing cables between components, and connecting speaker systems are explained. Systems are recommended for small, medium and large rooms. A list of steps for testing a sound system is included, both with and without an audience.

The book concludes with a chapter about diagnosing trouble, repairing, and maintaining the system, and a chapter on adding additional components like wireless microphones, listening aids, special effects generators, CD players and cassettes.

"Is This Thing On?" was written and published for the person responsible to provide a sound system, repair it if it goes bad, and keep it maintained so it will operate with good consistent quality. That was our goal; we hope we have reached it.

GMcC
MPI

Introducing Sound Systems 1

INTRODUCTION

Hearing is only one of the five human senses, and its importance has often been underrated because the sense of sight is so prominent. Time and again, statistics have shown that we favor pictures over sound. Learning is some eighty percent faster with pictures. Our brain responds more than twice as fast to something we see as compared to something we hear. And who ever said "my ears were bigger than my stomach" after trying to eat all of a triple-scoop banana split? Even so, the quest for good sound reproduction has almost become an obsession and continues headlong each day.

SOUND SYSTEM PROGRESS

The science of sonics has undergone a revolution during the past 20 or so years. It wasn't long ago that TV sets—even the luxury color models—were built with scrawny three- or four-inch speakers. Today, some television sets have an elaborate built-in sound system or have output jacks to connect to a stereo system for much livelier sound.

Years ago, movie theaters had just one speaker behind the screen. Now, movie theaters are equipped with ultra-modern sound systems that literally surround you with high-quality music and sound effects—gunfire, explosions, jet planes whooshing by—that make you feel like you are *in* the action on the screen.

For many years, home hi-fi fans worried about keeping their phonograph records clean and scratch free. Today, the photograph record is almost obsolete for new recordings since it has been replaced by the magnetic tape cassette and compact disc (CD).[1] A typical $300 home stereo system uses advanced components like CD players, digital signal processors (DSPs), and complex speaker systems to deliver a clean, crisp sound—sound quality that used to cost thousands of dollars.

The result of this audio revolution is that the average person—not just the audiophile—is now accustomed to true high-fidelity stereo sound where voices are clear and the full range of an orchestra is faithfully reproduced. The tinny, muffled squawks that were common from yesterday's sound systems are no longer acceptable.

Building a good sound system for the home or car—or improving on an existing one—is relatively straightforward, and costs can often be maintained within a specified budget. Specific home and car projects, completely designed with cabinet layouts, are presented in *Speakers for Your Home and Automobile*, G. McComb,

A. Evans, E. Evans, and *Advanced Speaker Designs*, R. Alden. These books are published by PROMPT Publications. However, it's not as easy to build or improve a public sound system—the kind you'd find in a church, synagogue, meeting hall, or conference room. Such systems are typically more complex than home or automobile systems because they must amplify and project the sound so that many people in a large room can hear it. Public sound systems require much more planning than personal sound systems, and they can be quite costly.

ABOUT THE BOOK

This book addresses the unique problems and challenges of the installation, use, and maintenance of public sound systems. It provides information on:

- Basics of acoustics
- Component parts of public sound systems
- How to read and understand sound system specifications
- Installing the parts of a sound system, including amplifiers and loudspeakers
- Upgrading an existing sound system
- How to operate a sound system
- Diagnosing, maintaining and repairing a sound system
- And more

This book is intended to be a comprehensive learning guide and operation manual for the people responsible for installing, using, and maintaining a public sound system used in a room with a maximum capacity of 1200 or fewer people (about the size of a large school gymnasium). Larger rooms and outdoor installations usually require special equipment and engineering knowledge that is beyond the scope of this book; so, discussions of sound systems for very large auditoriums, football stadiums, outdoor amphitheaters, and similar installations are not included.

WHAT IS A PUBLIC SOUND SYSTEM?

A public sound system comprises all the components used to amplify and distribute sound in a public or commercial environment. Public sound systems differ from "private" sound systems that you might find in a home. Home stereo systems are generally designed to reproduce sound for a few people in a small room. Public sound systems, on the other hand, are typically used to amplify signals that consist of live or prerecorded music and voice so that many people in a rather large room (often auditorium size) can hear them.

Sound Reinforcement Systems

Of the several sub-groups of public sound systems, the most common is the sound system in a large auditorium. Examples are churches, synagogues, lodges, meeting halls, school theaters, conference rooms, and gymnasiums. A sound system amplifies a sound signal from a source such as a microphone or tape player. One of the most basic systems is shown in *Figure 1-1*. It consists of a microphone, amplifier, and speaker system. The amplified signal is applied to large speaker systems that reproduce the input sound at loud volume. This type of public sound system is technically referred to as a sound reinforcement system, because the system "reinforces" the original sound and makes it loud enough so everyone in the room can hear. For this book, we will consider that *a sound reinforcement system is designed for amplifying speech or music for a single area or single building.* Throughout the remainder of this book, unless otherwise noted, we will refer to a sound reinforcement system simply as a "sound system."

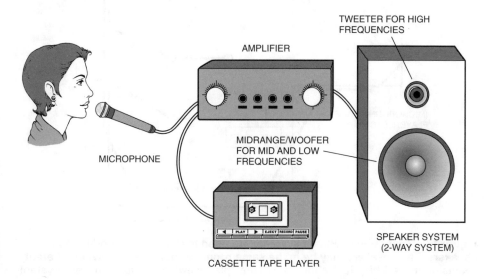

Figure 1-1. The most common components of a basic sound system are: microphone, amplifier, and speaker system. A common accessory is an audio tape recorder.

Public Address System

Another sub-group of public sound systems is the *public address system* (PA system). It also is often referred to as a sound reinforcement system, but there are some differences between the two. For the sake of clarity, in this book we will consider that *a public address system is designed primarily for amplifying speech through a network of two or more distant speaker systems*, which may or may not be in the same room or area. For example, as shown in *Figure 1-2,* such a public address system could be used to distribute sound throughout an office building. (Although such a system could be used within a school building, a school system is more often of the intercom sub-group which provides selection of rooms and usually is a two-way system; whereas, a public address system is a one-way system with no selection of rooms or areas — everybody within range hears it.)

Many of the basic components are the same for a sound system and a public address system; however, the planning and installation are often different. We will cover basic installation and maintenance of public address systems in later chapters, but we will concentrate our main focus on sound systems, as they are more likely to require special attention. For example, a public address system is generally installed by a building contractor and operation consists only of turning it on and adjusting the loudness, while a sound system is usually a special installation.

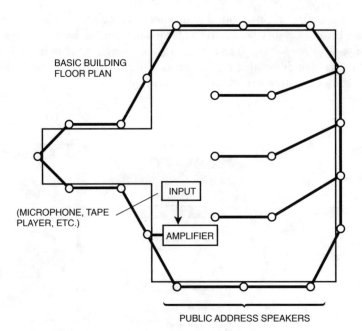

BASIC BUILDING
FLOOR PLAN

INPUT

(MICROPHONE, TAPE
PLAYER, ETC.)

AMPLIFIER

PUBLIC ADDRESS SPEAKERS

Figure 1-2. Most public address systems are used to distribute sound to many speakers. The speakers may be in the same room, or more commonly, in different rooms or areas. This illustration shows the "plan view" of a public address system installed in a small building.

BASIC COMPONENT PARTS

A sound system is comprised of three major parts as shown in *Figure 1-3:* an input transducer, a signal processor, and a speaker system. All sound systems are designed around this simple core of components, and differ only in the number, type, and quality of input transducers, signal processors, and speaker systems. However, the variations that are possible in sound system design are many — depending on the size and type of room, as well as the purpose of the sound system.

We will discuss these three major components in more detail later, but for now, let's summarize what each component does, and how it contributes to the overall function of the sound system.

BASIC SOUND SYSTEM

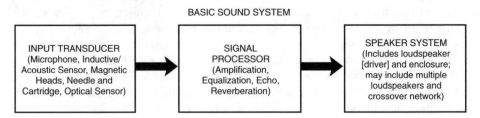

INPUT TRANSDUCER
(Microphone, Inductive/
Acoustic Sensor, Magnetic
Heads, Needle and
Cartridge, Optical Sensor)

SIGNAL
PROCESSOR
(Amplification,
Equalization, Echo,
Reverberation)

SPEAKER SYSTEM
(Includes loudspeaker
[driver] and enclosure;
may include multiple
loudspeakers and
crossover network)

Figure 1-3. Every sound system contains three basic elements: input transducer, signal processor, and speaker system.

Input Transducer

Input transducer is the general name for a device that converts a live or recorded sound into an electrical signal so that it can be amplified. Several kinds of input transducer can be used in a sound system.

The most common input transducer is the microphone, which converts air movement into an electrical signal. Many types of microphones have been developed to meet specific needs. Sensitivity, frequency range, and directional pattern are some of the parameters that need to be considered when selecting a microphone.

Another input transducer often used in sound systems is the inductive instrument pickup which is used with metal-stringed instruments such as an electric guitar. The inductive instrument pickup converts the motion (not the sound) of the metal strings into electrical signals for amplification. The acoustic instrument pickup, which is essentially a special type of microphone, attaches to the instrument to pick up the actual sound produced by the instrument. An acoustic pickup might be attached on or near the soundboard of a violin or acoustic guitar, or at the bell of a saxophone.

Also considered input transducers are the magnetic heads in a tape player, the needle and cartridge in a turntable (phonograph), and the optical pickup assembly in a compact disc player. All three of these input transducers convert stored sound signals into electrical signals so that they can be amplified. The way these input transducers operate is not important here; you just need to know that they are among the many types of input transducers you may encounter as part of a sound system and that their specifications may be important to the overall system performance.

Signal Processor

A sound system would be pointless if it didn't modify the sound in some way. The basic function of a sound system is to make the original sound source louder so that everyone in a large room can hear it. This requires amplification, which is the basic form of *signal processing*.

A signal processor can alter the sound in other ways. It is routinely used to increase or decrease the relative energy levels of various frequency bands to change the tonal rendition of the sound. This is most often done to tailor the sound to the room. This process is called *equalizing*, and takes its name for the function of "equalizing" the sound to match the unique acoustics of the room.

For music presentations, sound systems can provide signal processing which produces echo and reverberation. These acoustic special effects makes the sound source seem more "full." Another typical use of echo and reverberation is to enhance the sound of a singer, and is one of the techniques popular in Karaoke. With echo and reverberation, a singing voice seems richer and stronger.

Speaker System

No sound system is complete without a speaker system which converts electrical signals into sound waves. As used in this book, the term speaker system includes the loudspeaker or driver, the crossover network (if any), and the speaker enclosure. A typical speaker system for a sound system has two, three, or four loudspeakers inside an enclosure. Each loudspeaker reproduces a different portion of the sound range. For example, three loudspeakers are used in a 3-way system: one loudspeaker (called woofer) delivers low tones, the second (called midrange) delivers the midrange tones, and the third (called tweeter) delivers the high tones. Working together in one speaker system, and usually with the help of a crossover network, these individual speakers provide the full range of sound. A 2-way system is shown in *Figure 1-1*.

HOOKING IT ALL TOGETHER

Of course, the three major parts of a sound system — input transducer, signal processor, and speaker system — must be connected together to provide a path for the electrical signals. This is the domain of *cables*. Cables connect the microphone to the amplifier, and the amplifier to the speaker system. Examples are given in *Figure 1-4*. Cables are also used to connect other input sources — instrument pickups, CD players, etc. — to the amplifier. In fact, in even a modest sound system, there may be dozens of cables.

At this point, it is helpful to differentiate between the terms *cable* and *wire*. Both are used to route electrical signals from one point to another, using a metal conductor inside an insulated covering. While purists argue there are definite differences between a cable and a wire, not everyone agrees on what makes a cable a cable, and a wire a wire!

In this book, we will use "cable" to refer to two or more conductors wrapped in a protective jacket with at least one end terminated in an attached connector. The conductors may be insulated wire (individual or twisted pair), braided shield, or some other type. For example, a common type of coaxial cable has a single solid wire as the center conductor which is encircled by insulation, a braided shield (the second conductor), and a protective insulating outer cover. Two connector types commonly used in sound systems are the familiar RCA® phono plug and the special XLR connector used in pro-level audio equipment. We will review connector types in Chapter 5.

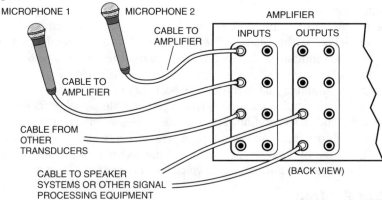

Figure 1-4. Cables are used extensively in sound system installation to connect the component parts together. Here, two microphones are connected to an amplifier via microphone cables, and an additional cable is shown going to another input source. Amplifier output cables go to speaker systems or other signal processing equipment.

A "wire" is a length of extruded copper or aluminum of various diameters. It may be solid or stranded. It may be bare, but more often is encapsulated inside an insulating cover. Although wire is commonly used for interconnections inside the major components of a sound system, it is not often used outside them. One exception is that sometimes an amplifier is connected to speaker systems by two insulated stranded wires that are bonded together. This type of wire is sometimes referred to as "zip cord" because the two wires can be easily separated at the "seam" between them.

BASIC CRITERIA FOR SOUND SYSTEMS

The design and complexity of a sound system is related to its purpose. A sound system used just for amplifying a speaking voice in a relatively small room — a conference room, for example — does not need to be as sophisticated or powerful as a sound system used for amplifying voice and music in a large auditorium.

While room size definitely plays an important role in determining the complexity of a sound system (we devote a section of Chapter 4 to this issue), the first criterion to be considered is whether the sound system is designed for voice only, or voice and music. Voice-only sound systems do not generally require the same level of equipment and planning as sound systems for voice and music. Knowing the kind of input signals that the sound system must amplify helps you prepare for the level of effort required to adequately plan, install, and operate the sound system.

When you plan a sound system, you must know the sound range of the input signals. Chapter 2 delves more deeply into what makes up sound, and defines the concept of sound range (more accurately called frequency range). But for the time being, you can think of sound range as the keys on a piano as shown in *Figure 1-5.* Keys on the left side are the low tones; keys on the right side are the high tones.

Human voices, especially speaking voices, are contained in a much smaller range of the piano keyboard. In fact, the range of a speaking voice is actually contained within only about 10 or 12 keys on the keyboard, somewhere around the octave below middle C. A sound system that deals only with voice needs only to amplify these tones; therefore, the sound system will be less complex.

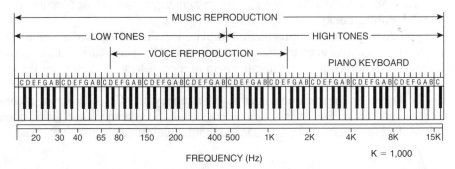

Figure 1-5. The range of sounds that must be reproduced is one of the key factors in determining the complexity of a sound system. Reproducing voice only is not as demanding as reproducing music, because the human voice is limited to a fairly small section of the sound range.

Conversely, a sound system for music must adequately amplify all of the frequency range of the piano keyboard. This requires better microphones (and usually more than one), higher quality and more powerful amplifiers, and speaker systems that can handle the frequency range and transients. The type of music amplified by the sound system also determines the complexity of a sound system. As a group, stringed instruments play the widest range of tones and a sound system used to amplify an orchestra containing violins, cellos, and harps is harder to plan, install, and operate than a sound system for only a tenor saxophone.

Chapter 3 and Chapter 4 provide more detail on matching a sound system to its intended use.

PRACTICAL KNOWLEDGE

Obviously, sound systems run the gamut from simple one microphone and loudspeaker jobs for portable Karaoke shows, to huge concert halls with an audience of several thousand people. The larger the sound system, the more specialized knowledge you need to install and operate it.

If you're involved in developing and maintaining a sound system for a church, meeting hall, school auditorium or similar room, the sound system probably will be of moderate complexity and you will be able to purchase components locally through specialty electronics outlets. To install and operate such a system, you don't need formal education in electronics or electricity. In most cases, you do not need a certificate, license or permit, and you are not required to pass an exam. (However, you should check your local and state codes to be sure.)

All you really need to successfully plan, install, operate, and maintain a moderate-sized sound system are basic technical skills. You should be comfortable with technical information and should have manual dexterity and experience to operate mechanical and electronic devices. (In other words, if you're a "techno-klutz," sound systems probably aren't for you.)

Prior knowledge of electronics and electricity is not required, though it is highly recommended. This book cannot address all of the fundamentals of electronics and electricity, therefore, we assume a certain level of basic expertise. If you would like to know more about electricity and electronics, you may want to purchase the book *Basic Solid-State Electronics*, Van Valkenburgh, Nooger & Neville, Inc., PROMPT Publications.

If you do not have the skills needed to plan, install, and operate a sound system, and don't have an acquaintance with the interest or qualifications that can do it for you, you are better off getting the help of a professional. You want the job done right the first time. A poorly planned or installed sound system can be a major headache, not to mention something of a money pit. Consider hiring a consultant or contractor. But even if you do not install the system yourself, this book will be valuable to help you better understand the requirements and limitations of your system.

TOOLS NEEDED FOR THE JOB

With only a couple of exceptions, no special tools are required to install, operate, or maintain a typical sound system. An assortment of basic tools, including screw drivers, wrenches, and pliers, will form the core part of your sound system toolbox. You will also want an assortment of hand tools for working with electrical wiring, including wire cutters, insulation strippers, and wire crimpers. Typical types are shown in *Figure 1-6*.

| a. Wire Cutter | b. Insulation Strippers | c. Wire Crimpers |

Figure 1-6. You need these three wire tools in your toolbox: a. wire cutter, b. insulation strippers, and c. wire crimpers.

We suggest you add the following specialty items to your sound system toolbox if you do not already own them. Both can usually be found in specialty electronics outlets for under $40 each. (Hands-on use of these tools is covered in later chapters.)

- ■Volt-Ohm-Milliammeter (VOM) or Multitester. A typical unit is shown in *Figure 1-7*. The "Volt" function is used to measure the dc or ac voltage at a point in a circuit. The "Ohm" function is used to measure electrical resistance of a wire or circuit component, and to check the continuity of a circuit. Some VOMs have a separate "Continuity Check" function with an audible or visual indicator of continuity. For example, you can use the continuity check to determine if a wire is broken. The "Milliammeter" function is used to measure circuit current. Usually the values are less than one ampere, but some have a 10A range. (Some VOMs can perform other tests, but these functions are not typically used in installing and operating a sound system.)
- ■Sound Level Meter. A typical unit is shown in *Figure 1-8*. A sound level meter allows you to measure the actual loudness and adjust the actual loudness of sound in a room. The meter has numerous uses, including helping you find "dead spots" in the acoustics of a room. (If for no other reason than for safety *and* liability, you should use a sound level meter to ensure that the sound level is not too high.)

SAFETY FIRST

Sound systems are relatively safe, and the latest components follow strict safety guidelines by both the manufacturer and the Underwriter's Laboratory (UL). However, care must still be exercised when installing, operating, and maintaining a sound system. Most sound systems are powered by 117 volts ac. A malfunctioning component or faulty wiring could expose someone to a shock hazard, so always use caution. This is especially important if the sound system is used outdoors, regardless of weather conditions.

Figure 1-7. If you plan to be serious about your sound system installation, you should have a basic VOM. (Courtesy of Radio Shack.)

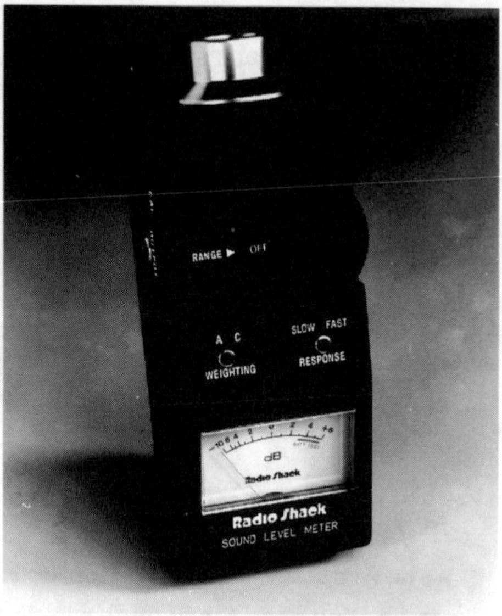

Figure 1-8. A sound level meter allows you to test the relative volume of sounds in a room. It is a handy tool to have when installing, using, and maintaining a sound system. (Courtesy of Radio Shack.)

Though *extremely* rare, a faulty or improperly wired sound system could cause serious injury or death to anyone touching one of the components, including a microphone. Chapter 5 includes some simple tests you can perform to ensure that your sound system is operating properly and is safe to use. You will need a VOM to conduct these tests. Start with these tests before you do any maintenance or repair of a sound system.

Unless you have specific training and experience, you should *never* open the case or cabinet of an ac-powered sound system component. Doing so may expose you and others to potentially lethal voltages. Proceed with caution.

SUMMARY

In this chapter you learned what a sound system is, and how it is used. And you learned that a basic criterion for the complexity of the sound system is the type of sound the system amplifies — whether voice only or voice and music. You also learned the practical knowledge and skills you need to successfully install, operate, and maintain a sound system, as well as the tools you should have for the job.

MOVING ON

There is, of course, much more to sound systems than we've covered here. In Chapter 2, you will learn the basics of the science of acoustics, including what makes up sound. You will also be introduced to a number of basic technical terms and concepts that will help you better navigate the sometimes rough seas of sound systems — including the difference between a dB and a dBm, how the inverse square law affects sound listening, and how room designs affect what we hear.

Learning the Science of Acoustics

2

INTRODUCTION

As we discussed in Chapter 1, the primary goal of a sound system is to make sound loud enough so that everyone in a room or auditorium can hear it. The amount of amplification a sound system uses to achieve this goal depends on the original sound source.

Some sound systems — like those used at a rock concert — are designed to amplify the original sound by a significant amount. The audience hears the band *much* louder than the instruments or singers really are. Other sound systems amplify the sound only enough so that those in the back of the audience hear the original sound as loudly as those in the front of the audience. This kind of sound system is typically used for voice, such as amplifying a lecturer in a meeting hall, or the minister in a church.

But amplification alone doesn't make a good sound system. The amplified sound has to be intelligible. This is where the design of a good sound system becomes a challenge. That means the sound can't be too "boomy," or else the audience won't understand the voices, or enjoy the music. Nor can the sound be filled with squeals, hum, or other noise. The amplified voice or music should sound as much like the original as possible.

One important consideration often left out of designing a good sound system for an installation is the room acoustics. You don't need to be a sound expert, but to assure your sound system will perform better, you need to take the time to learn some basics about the science of acoustics given in this chapter. You'll learn what makes up sound and how sound is affected by an enclosed room. You'll also learn some basic sound system terminology.

WHAT SOUND IS

On a broad level, the sound you hear is air molecules wafting toward and away from you. Sound starts at a source, for example, a person's mouth. The voice causes the air around the person to vibrate. This vibration is in a specific pattern called a wave, and this wave is not unlike the undulating ripples of a pebble thrown into a calm lake.

As the sound travels on this wave, it causes the surrounding air to alternatively compress and expand. That is, at one instant the air molecules are pressed tightly together, and at another instant, the air molecules are spreading apart. This change of pressure — compression and expansion — is what causes our eardrums to

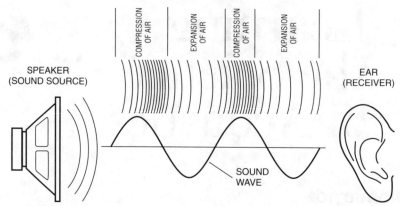

Figure 2-1. Sound is passed through air by alternately compressing and expanding pressure. Our eardrums move in response to the pressure changes, and this movement is transmitted as nerve impulses to the brain which perceives them as sound.

vibrate when the sound wave reaches us, as illustrated in *Figure 2-1*. The ear drum is pushed in on the expansion phase of the wave, and pulled out on the compression phase.

Sound Volume

As shown in *Figure 2-2*, the amount of air that moves in the wave largely determines the loudness of the sound. If you whisper, the whisper causes only a small amount of air to vibrate, so the volume is low. If you shout, the shout causes a lot of air to vibrate, so the volume is high.

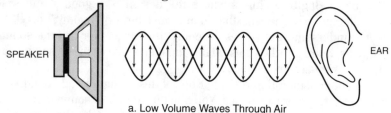

a. Low Volume Waves Through Air

Double arrow indicates up and down motion of air particles that propagate the sound wave from source to receiver. The particles remain in place; the wave propagates. Larger amplitude indicates louder sound.

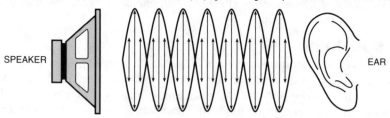

b. High Volume Waves Through Air

Figure 2-2. The changing pressure of a small amount of air is the result of quiet sounds. Conversely, the changing pressure of a large amount of air is the result of louder sounds.

Source: *Speakers for Your Home and Automobile*; G. McComb, A.J. Evans, E.J. Evans; ©Copyright 1988, 1991, 1992, Master Publishing, Inc. Published by PROMPT Publications.

Volume is a poor term to use when defining the loudness of sound, because the term is inexact. While "volume" is certainly an acceptable word (and we'll use it from time to time throughout this book), a better term that describes the loudness of sound is *sound pressure level*, or *SPL*. SPL is commonly expressed in decibels, a logarithmic unit of measure. The decibel (dB) is not an absolute measure, but rather it indicates a ratio between two sound levels. Zero dB represents the SPL of a certain tone that can just be heard by a person with normal hearing. The level of other sounds is measured or expressed relative to this lowest audible sound level. For a person of normal hearing, a change in SPL of 10 dB is perceived as being twice as loud as the original sound. *Figure 2-3* shows a graph depicting various sound intensities and their relative levels in dB.

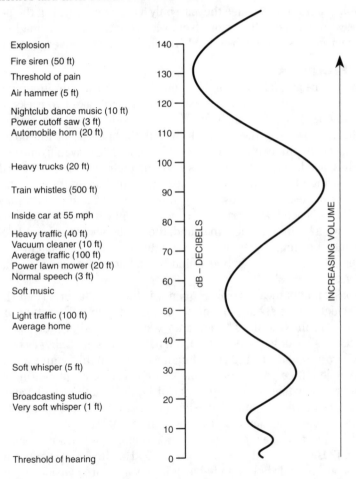

Figure 2-3. Various sound levels in dB SPL. Soft music (as heard near the speaker of a home hi-fi system, for example) is approximately 60 dB SPL. Loud music is about 105 to 115 dB SPL. The threshold of pain for most humans is about 130 dB SPL.
Source: *Speakers for Your Home and Automobile*; G. McComb, A.J. Evans, E.J. Evans; ©Copyright 1988, 1991, 1992, Master Publishing, Inc. Published by PROMPT Publications.

Sound and the Inverse Square Law

Look again at *Figure 2-3*. The measurements from the various sound sources assumes a certain distance as indicated in the figure. You know from experience that a fire engine siren sounds much louder at a distance of 50 feet than at 1,000 feet.

This demonstrates the *inverse square law*, which states that the level of sound decreases 50 percent for every doubling of the distance from the sound source. For example, normal speech has an SPL of about 62 dB measured at three feet. Step back from the speaker so you are six feet away and the SPL will be about 31 dB. Walk out 12 feet and the SPL will be about 15.5 dB, and so forth.

The decrease in sound level with distance is the reason a sound system is needed in an auditorium. The people in the front row can hear the speaker's natural voice quite well because they are only a few feet away. But the people in the back row, who may be 100 or more feet away from the speaker, might hear only a faint whisper because of the drop in SPL as defined by the inverse square law.

Sound Frequency

Next time you're at a lake or pond (a pool or bathtub will do in a pinch), try this simple experiment. With the water calm and smooth, take two rocks, a small one and a large one. Drop the small rock over one side of the boat and observe the wave pattern. Now drop the big rock over the other side of the boat. The wave pattern caused by the two rocks will be different. The waves from the small rock are closely spaced, whereas the waves from the big rock are spaced further apart.

The spacing of the waves determines the frequency of the wave pattern. This frequency is a measure of how many waves go past a stationary object in one second. Waves spaced far apart have a low frequency, whereas waves spaced close together have a high frequency. In sound, the frequency of the sound determines if it's low pitched or high pitched. As you might surmise, a low-pitched sound has a low frequency, and a high-pitched sound has a higher frequency, as illustrated in *Figure 2-4*.

The basic unit of measure for frequency is the *hertz* (abbreviated Hz), after the German scientist who spent most of his lifetime studying the nature of waves and frequencies. (In the "old days," frequency was expressed in cycles per second, or cps; this term is no longer widely used, but means the same thing as hertz). For the relatively low frequency of 100 Hz, a hundred waves of this sound will travel past a stationary object (like your ears) in one second. You sense of hearing will register this sound as a low tone. For a frequency of 1,000 Hz, a thousand waves will travel past your ears in one second. Your sense of hearing will register this sound as a much higher tone — in fact, ten times as high as 100 Hz.

Humans hear in a relatively narrow range of frequencies from about 20 Hz to 20,000 Hz. (20,000 Hz is often expressed as 20 kHz; the *k* represents a multiplier of one thousand). Sounds below and above these values are considered subsonic and supersonic, respectively. The ability to hear very low and very high frequencies depends on age, sex, and prior exposure to loud sounds. Generally speaking, many men past 40 have difficulty hearing sounds above 15,000 Hz, whereas some teenagers (male or female) might be able to hear sounds in excess of 20,000 Hz. Although hearing range varies, it is generally accepted that the "normal" range is 20 Hz to 20 kHz in sound system designs.

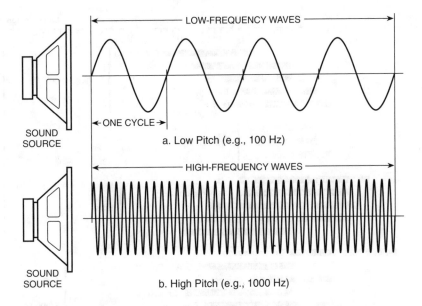

Figure 2-4. The pitch of a sound is determined by its frequency, which is the number of waves that pass a stationary object in one second. A low-pitched sound has a low frequency; a high-pitched sound has a high frequency.

Frequency range is an important aspect of sound systems. As illustrated in *Figure 2-5,* the human voice and musical instruments produce sound over a wide range of frequencies. Our ears are used to hearing these sounds at their normal frequencies. This means a sound system that cannot adequately reproduce these frequencies will not sound natural.

Three Ranges of Sounds

The audio spectrum of 20 Hz to 20 kHz can be subdivided into three convenient categories: low pitch (low frequency), medium pitch (medium frequency), and high pitch (high frequency), as indicated in *Figure 2-6*. These categories are often used instead of the actually frequency in hertz to describe the relative strengths or weaknesses of a given sound system.

- Low-pitch sounds, referred to as *bass*, are made by bass singing voices and by bass instruments, such as kettle drums, tubs, bassoon, and string bass.
- Medium-pitch sounds, called *midrange*, are made by most singing voices, guitars, and most other musical instruments. Because most sounds are in this range, it is the most important.
- High-pitch sounds, referred to as *treble*, are made by bells, cymbals, flutes, violins, and crickets trapped in the sound studio.

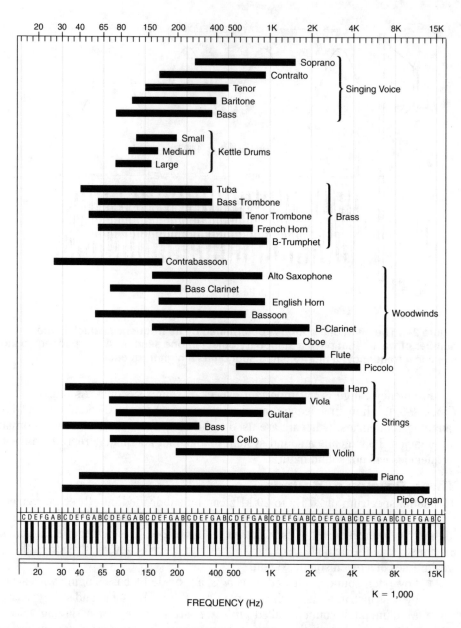

Figure 2-5. The range of human hearing extends from about 20 Hz to 20 kHz. The human voice and musical instruments fall within this range. Note that many instruments, particularly the string variety, span a wide range, and require a sound system that accurately reproduces these frequencies.

Source: *Speakers for Your Home and Automobile*; G. McComb, A.J. Evans, E.J. Evans; ©Copyright 1988, 1991, 1992, Master Publishing, Inc. Published by PROMPT Publications.

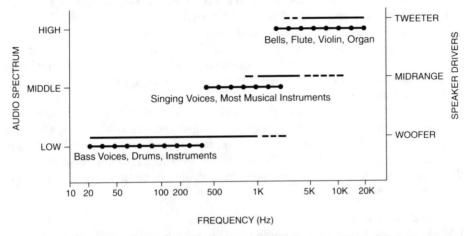

Figure 2-6. The audio range of human hearing may be divided into three groups: low, middle, and high frequencies. Different loudspeaker drivers are designed to cover the frequency spectrum.
Source: *Speakers for Your Home and Automobile*; G. McComb, A.J. Evans, E.J. Evans; ©Copyright 1988, 1991, 1992, Master Publishing, Inc. Published by PROMPT Publications.

Sound Frequency vs. the Speed of Sound

A common mistake is to confuse the frequency of sound with the speed of sound. Air pressure and temperature affect the speed of sound, but at a given combination of these, the speed is constant regardless of frequency. The speed of sound is defined as 1,130 feet per second at normal air pressure at sea level at an air temperature of 59 degrees Fahrenheit.

Because the speed of sound is finite, and is actually rather slow, it can be a problem in some sound systems. In a very large room (or an outdoor stadium), listeners far away from the speakers may hear the sound a third- to a half-second later than people close to the speakers. To prevent this, and to make sure the sound is loud enough in the back of the room, professional sound system designers often place auxiliary speakers at the distant areas. Because the people in the back of the room are close to the auxiliary speakers, they hear the sound without delay.

However, auxiliary speakers can cause another problem — echo — unless the sound system is very carefully engineered. People in the back will first hear the sound from the nearby auxiliary speakers, then hear the same sound from the distant main speakers. When auxiliary speakers are used in large, complex sound systems, an electronic circuit may be used to delay the signals going to the auxiliary speakers so that the sound from the auxiliary speakers arrives at the people's ears at the same time as the sound from the distant speaker. The people in the back of the room still hear the sound delayed, but without echo.

Sound systems with signal delay are beyond the scope of this book. If you are in charge of installing or using a sound system for a very large area (over 250-300 feet deep), you may need to consult with a sound system expert on the use of auxiliary speakers and/or signal delay processors.

HOW SOUND IS AFFECTED BY AN ENCLOSED ROOM

The sound you hear is greatly affected by its environment. This is most obvious in a small, enclosed area with smooth, hard-surfaced walls. For example, the sound of footsteps is completely different in the open air than in an enclosed room. In the open air, you hear just the actual sounds of your feet stomping on the ground. But in the enclosed room, you hear the sounds of the footsteps, plus numerous echoes as the sound bounces from wall to wall.

For sound systems that are used indoors, the size, shape, and other factors of the room greatly determine the overall quality of sound. There are two major ways enclosed rooms affect sound: *reverberation*, and its converse, *absorption*. The following paragraphs introduce the concepts of reverberation and absorption, and their causes. Chapter 4 details how you can control reverberation and absorption to achieve better sound.

Reverberation

Just like the walls of the closed room cited above, boundaries of a room will cause some echoes of the original sound. These echoes — more accurately called reverberations — may be insignificant, or they may be so overpowering that people hear more echo than original sound!

Reverberation is caused by sound bouncing off boundaries of a room. These boundaries include walls, floor, ceiling, and any other object that is large and hard enough to reflect sound.

Contrary to popular belief, the "ideal" room is not reverberation-free. In fact, our ears and brain are trained for a certain amount of reverberation, so a completely echo-free sound may seem unnatural. Reverberation is part of normal sound, and only excessive reverberation is considered detrimental to a sound system.

A reflected sound is not as loud as the direct sound that produced the reflection. The relative loudness of the reflected sound depends on a number of factors: how much sound is reflected from the boundary, the angle of the boundary in relation to the sound source and the listener, and whether the source of the reflected sound is itself a reflection. In a simple square-shaped room, as shown in *Figure 2-7*, the main sound source in the center of the room is reflected many times, with reflections causing other reflections. While these reflections may appear to make a muddle of the sound, as long as the reverberation is not overpowering, the echoes will be perceived by the sense of hearing as natural and acceptable.

Absorption

In a typical room filled with an audience, the bulk of the sound is absorbed rather than reflected. The "hardness" of the boundary largely determines how much of the sound is absorbed. Hard boundaries, like a brick or smooth plaster wall, absorb very little sound. That means most of the sound is reflected, and is heard as reverberation.

Conversely, soft boundaries, like carpeting, acoustic tile, drapes, and even people, can absorb quite a bit of sound. As more sound is absorbed, less sound is reflected, which reduces the amount of reverberation.

Table 2-1 shows the relative absorption coefficients at 1 kHz of different kinds of common boundaries. An absorption coefficient of 0 means total reflection and an

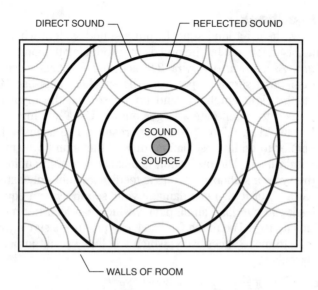

Figure 2-7. Most reverberation within a room is caused by sound bouncing off walls, ceilings, and floors. A smooth, hard-surfaced, solid object larger than a square foot may also cause sound reflections.

absorption coefficient of 1 means total absorption. Values in between are shown as a decimal fraction, making it easy to calculate percentage. For example, acoustic tile has an absorption coefficient of 0.70, so 0.70 times 100 = 70%. This means that 70% of the sound directed at acoustic tile is absorbed by the tile.

In almost all cases, absorption of a given material decreases with lower frequencies and increases with higher frequencies. Since more of the high frequencies are absorbed, the lower frequencies sound louder, even if the sound equipment is reproducing all frequencies equally. This is why the sound may be overly "boomy" or "bassy" in some rooms. You will learn in Chapters 4 and 7 how to use absorbing materials to adjust the system so the sound is more pleasing and natural.

Table 2-1. Absorption coefficients of selected boundary materials.

Material	Absorption Coefficient
Acoustic tile	.70
Audience	.95
Brick wall, painted	.02
Brick wall, unpainted	.04
Carpeting	.30
Drapes	.75
Plaster wall	.06
Plywood panel	.20
Poured concrete	.02
Upholstered seats	.90
Wood flooring	.08

Acoustic Feedback

You probably have heard a loud squeal from a sound system when the amplifier volume was increased too much. The process of producing the loud squeal is called *acoustic feedback.* As shown in *Figure 2-8,* acoustic feedback occurs because a portion of the amplified signal output from the speakers gets back to the microphone with just the right amplitude and timing so that the input to the sound system is reinforced to regenerate a larger output. Usually a high-pitched noise, often referred to as "ringing" or "squealing" results. The acoustic feedback has caused the amplifier to go into oscillation; that is, the sound system has regenerated a signal within itself. The oscillation is usually at a high audio frequency — one at which the system stabilizes. Acoustic feedback is not only extremely irritating to your hearing, but it also can damage your sound system equipment. High levels of feedback can overload an amplifier, which can lead to premature failure (or at best, a blown fuse). Too much feedback can produce a signal that causes the loudspeaker cone to move excessively, which will likely result in permanent damage to the loudspeaker.

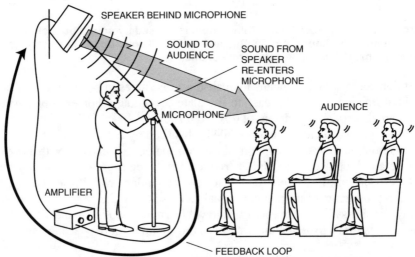

Figure 2-8. Feedback occurs when a certain amount of amplified signal re-enters the microphone of the sound system. It causes the amplifier to oscillate and generate a high-pitched squeal.

Preventing Acoustic Feedback

The most common way to avoid feedback is to turn down the output level ("volume") on the amplifier. This reduces the output from the speakers, so less sound reaches the microphone. However, this is not always the most useful technique, since reducing the amplification is counter to the reasons for using a sound system in the first place. In addition, intermittent squealing can still occur if the output level is reduced to just below the feedback point. Loud sounds, such as a cymbal crash, can cause momentary feedback. To stop this feedback, it is often necessary to either turn off the microphone or adjust the amplifier output level even lower. Obviously, this is not an acceptable solution for most sound presentations, especially music.

As shown in *Figure 2-9*, a proven way to minimize feedback is to place the speaker systems so that none point directly at the microphone and to use a directional microphone; that is, one that picks up more sound directly in front of it rather than any sound from behind it. Placing speakers behind a live microphone is a "no-no" for sound systems. In the typical sound system installation, the speakers are placed between the microphone and audience so that the small amount of amplified sound that re-enters the microphone is not enough reinforcement to cause feedback. Chapter 3 provides details on directional microphones. Chapter 4 provides additional hands-on guidance for proper microphone and speaker placement to prevent feedback.

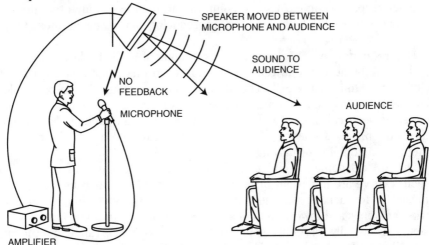

Figure 2-9. Feedback can be effectively controlled by placing the speakers between the microphone and audience.

BASIC SOUND SYSTEM TERMINOLOGY

No study of sound systems would be complete without a short introduction to the unique terms and phrases that are commonly used. We've already introduced six of the most common terms: sound pressure level, decibel, frequency range, reverberation, absorption, and feedback. Here are several other terms you need to be acquainted with as you learn about sound systems.

Frequency Response

Although the span of human hearing is 20 Hz to 20 kHz, most sound systems cannot adequately reproduce this entire frequency range. The range of frequencies that can be faithfully reproduced by the sound system is known as its *frequency response*. For example, a given sound system may be incapable of reproducing sounds below 100 Hz and above 15 kHz. Therefore, the frequency response of this sound system is said to be 100 Hz to 15 kHz.

At first glance, it may appear that frequency response is a fairly simple measurement of the upper and lower limits of sound reproduction that the system can provide. But this is not a complete picture of frequency response. The upper and lower limits of sound reproduction aren't brick walls. It's not as if a sound system with a frequency response of 100 Hz to 15 kHz cannot reproduce *any* sound below

100 Hz or above 15 kHz. Usually, the sound system is capable of reproducing some sound beyond these ranges, but the sound level is greatly diminished.

A graph of a typical frequency response is shown in *Figure 2-10*. The curved line represents the frequency response of a sound system; that is, the measured output at various frequencies with a constant input level. Notice that 0 dB is the *reference level*, not an absolute level, for the measurements. If the frequency response curve goes above the 0 dB point, it means the sound system puts out more sound at these frequencies than it does at the reference frequency, which is often 1 kHz. If the curve goes below the 0 dB point, it means the sound system puts out less sound at these frequencies. Notice the drop off in output at each end of the frequency response curve. At some point, the drop off is so much that the sound system is considered to be incapable of reproducing sound beyond these frequencies.

The frequency response graph shows the output in dB to denote the relative signal strengths at different frequencies. Earlier in this chapter you learned that the dB is used to measure SPL. The dB is also commonly used to measure electrical signal levels in electronic circuits, however, the two uses are quite different. In the case of electrical signals, a change of +3 dB means a doubling of signal power and −3 dB means a halving of signal power. Recall that a +10 dB or −10 dB change was a doubling or halving, respectively, of SPL.

The −3 dB points, f_1 and f_2 in *Figure 2-10,* at each end of the frequency response curve are important because they are often used as reference points to compare the frequency response of a sound system. By convention, a drop of 3 dB marks the upper and lower limits of sound reproduction. In other words, a sound system may actually reproduce sound over a 20 Hz to 20 kHz range, but the output is −6 dB at 20 Hz, −3dB at 40 Hz, −3 dB at 15 kHz and −10 dB at 20 kHz. As a result, the system is specified as having a 40 Hz to 15 kHz frequency response.

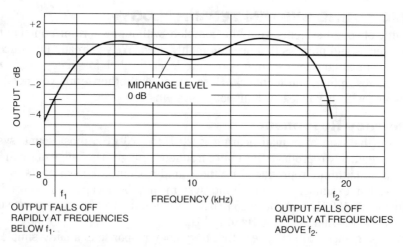

Figure 2-10. Typical frequency response curve for a sound system. The x-axis of the graph is marked in Hz to denote the frequency of sound. The y-axis of the graph is marked in dB, with 0 dB being the reference output level.

Source: *Speakers for Your Home and Automobile*; G. McComb, A.J. Evans, E.J. Evans; ©Copyright 1988, 1991, 1992, Master Publishing, Inc. Published by PROMPT Publications.

Dynamic Range

Dynamic range is the difference between the softest and the loudest sounds a sound system can faithfully reproduce. Even the best sound system is incapable of reproducing both the sound of a pin dropping and the blast of a cannon, and everything else in between! Many sound systems are designed to reproduce the sound levels between a whisper and moderate shouting. This is actually a fairly wide dynamic range, more than 50 dB SPL.

Often, a sound system is called on to reproduce a greater range of sounds. For example, a rock concert may require a dynamic range of 90 dB SPL to adequately reproduce all the music and singing. A 90 dB SPL dynamic range is actually quite a large span, and many sound systems have difficulty achieving it. For this reason, a common feature of sound systems is the "compression" circuit. As its name implies, the circuit compresses a wide dynamic range into a smaller range, so that the sound system can adequately reproduce it. All the sounds remain, but the loudest and softest passages are compressed closer together. You'll learn more about compression in later chapters.

Headroom

Headroom is related to dynamic range. Headroom is the amount of dynamic range between the average sound pressure level and the absolute maximum sound level as shown in *Figure 2-11.* The greater the headroom, the higher the "peaks" that the sound system can accommodate. A sound system that does not have adequate headroom will not sound natural, because it won't be able to handle louder-than-normal passages of a program. You will learn in subsequent chapters that headroom is something that you can plan for in a sound system design, and that it can be controlled if your sound system employs a compression circuit.

Noise Floor

The *noise floor* is ambient sound and system noise below the program sound. For example, suppose the noise floor is at 30 dB, which, as shown in *Figure 2-3,* is the sound level of a soft whisper at five feet. Any sounds softer than this will not be adequately handled by the sound system because the sounds are below the noise floor. Note that the noise floor also includes system noise, which is the hiss and/or hum you hear when you turn up the output level of an amplifier. The hiss is random noise generated by the electronic components of the system and can never be completely eliminated. Hum comes from incomplete filtering in the equipment power supply; it is also difficult to completely eliminate. Better sound systems have less noise, but some noise is always present.

As shown in *Figure 2-11,* the difference between the noise floor and the maximum sound pressure level denotes the dynamic range. Notice that sound below the noise floor is not considered part of the dynamic range of a system, therefore, because there is always some noise in a sound system, dynamic range never starts at 0 dB SPL. *Figure 2-11* also shows that the range between the noise floor and the average sound level is used to determine the signal-to-noise ratio. Notice that this range does not include the headroom.

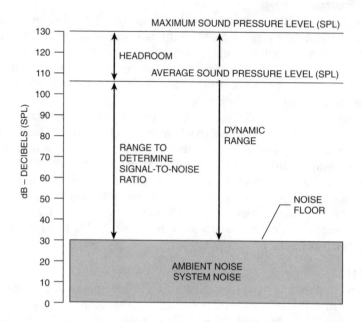

Figure 2-11. Dynamic range is the ability of a sound system to adequately reproduce both soft and loud passages. Dynamic range involves average sound pressure levels, maximum sound pressure levels, noise floor, and signal-to-noise ratio.

Various Forms of dB Measurements

Recall that the decibel (dB) is used to denote a ratio. It's used in SPL measurements to indicate the difference in levels of sounds from a variety of sources. Zero dB is used to denote the threshold of hearing. Soft music has a sound pressure level of about 60 dB, while a rock concert might have a sound pressure level of 115 dB. The decibel makes comparisons between sound pressure levels quite easy. For example, the difference in sound pressure level between soft music and a rock concert is 55 dB.

As explained previously, the decibel is also used for specifying frequency response of a sound system. Though both frequency response and sound pressure level measurements use dB, exactly what they measure are completely different. One measures sound pressure level and the other measures electrical signal strength. For this reason, you must avoid the temptation to mix dB measurements. The 0 dB line in a frequency response graph has no bearing to the 0 dB used in sound pressure levels to denote the threshold of hearing.

While measuring everything in decibels may be a bit confusing at first, the dB is a handy tool, and you should take the time to get to know what it means and what it represents. An important facet of dB measurement is that the decibel is based in a logarithmic scale, rather than a linear scale. This is ideal because the sensitivity of ours ears is also logarithmic.

By itself, a decibel measurement doesn't tell you much. For example, if some-one tells you the level of a sound is 3 dB, does he mean sound pressure level (SPL) or electrical signal strength? Remember that the two mean completely different things. Also, because decibels only describe ratios, and not absolute measure-ments, for decibels to be useful you must be told the reference value. For SPL measurements, the reference value of 0 dB for the threshold of hearing is inferred. But for signal strength, the reference value is nebulous, because it differs depend-ing on what is being measured. Is the reference value one volt? One watt? Or something else? The reference value can be important, because it helps you deter-mine what is being measured.

In an effort to help simplify and clarify the use of dB measurements, several standards have been developed through the years. The most common for sound systems is the *dBm*, which is measured with a specific reference value of one milliwatt (1 mW or 0.001W) of power applied to a circuit with an impedance of 600 ohms. (Impedance is the measure of opposition to current in an electrical circuit and is expressed in ohms or k ohms, where *k* stands for one thousand). The strength of the signal that results in 1 mW into 600 ohms is considered 0 dBm as indicated in *Figure 2-12*. All other measurements are based on this reference value.

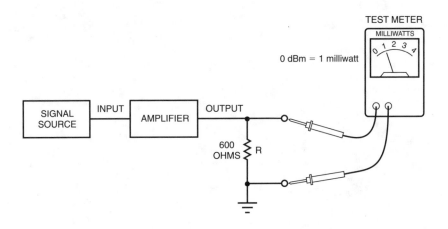

Figure 2-12. With the exception of the power amplifier, the output of most sound system components is measured in dBm, where "0 dBm equals one milliwatt into 600 ohms" is used as a reference value for other measurements. A typical test setup to measure 0 dBm is shown.

You may see dBm used in some sound system specifications, which is why it's important to understand what it means. You might also see the functional equiva-lent of dBm using just the standard dB measurement, along with a notation that the test was referenced at one milliwatt. For example, the output of a sound system mixer console might be listed as "+25 dBm" or "+25 dB above 1 milliwatt." Both mean the same thing.

There are other reference values used with dB, but they are not as common.

- *dBu* — dBu is similar to dBm, but with dBu, the impedance of the reference input circuit is not a factor.
- *dBW* — dBW is used when measuring large power outputs. 0 dBW equals one watt instead of one milliwatt.
- *dBV* — The dBm and dBu use power (milliwatts) as the reference value, but dBV uses voltage as the reference value. 0 dBV equals 0.775 volt rms. (rms means "root mean square," and is the usual unit of measure for ac voltage.)
- *dB PWL* — The dB PWL is another way of expressing acoustic power of a sound system. The dB SPL measurement described earlier in this chapter is more commonly used to measure acoustic power.

USING WHAT YOU KNOW

You are now ready to apply what you have learned about the basics of acoustics as well as the terminology used in sound systems to select, install, use, and maintain your own sound system. The next chapter discuses specifications you are likely to encounter when reviewing sound systems, whether you are buying a new system or upgrading an old one. An overview of specifications for sound system amplifiers, microphones, mixing consoles, and speakers is included.

Understanding Sound Systems Specifications ③

INTRODUCTION

Sound may be a natural phenomenon, but "natural" doesn't always mean simple. As you found in the previous two chapters, sound is a complex science. Numerous factors influence how we hear — whether a sound is soft or loud, smooth or raspy, breezy or boomy.

In the perfect world, a sound system would reproduce a signal as an exact duplicate of the original source. Put in an E-flat note from a clarinet, for example, and out comes an E-flat note with a sound that is indistinguishable from the real clarinet. But we don't live in a perfect world. Every element of a sound system — and we do mean *every* — "colors" (or alters) the sound in one way or another. This coloration may be subtle or it may be obvious. That E-flat note may sound genuine except for a slight warbling that only a musicologist would notice. Or the E-flat note may come out sounding more like a kazoo than a clarinet.

Technical specifications are used in an effort to quantify the way sound is altered by a sound system component. Some of these specifications are plain and obvious, while others are a bit more indistinct. Specifications are important for two reasons:

- The more you know about the behavior of your existing sound system components, the better you will be able to compensate for any deficiencies. The specifications help you evaluate how each individual component affects the sound system.
- Should you be in the market for new sound system components, you will be able to spot quality products by comparing their specifications.

In this chapter, we review the most commonly used sound system specifications for each of the four main components: microphone, mixing console, power amplifier, and speaker system. If you are familiar with hi-fi technology, many of these specifications will already be familiar to you, and you may wish to skim through this chapter, and go on to the next one.

Review of a Typical System

Chapter 1 introduced the basic components of a sound system. In a real world system, you are likely to encounter a sound system made of the following, as shown in *Figure 3-1,* which is an expansion of *Figure 1-1.*

- One or more microphones, with each microphone positioned in front of a person or instrument.

- A mixing console combines two or more sound sources. These sources can be microphones, as well as CD players, tape decks, and turntables.
- A power amplifier boosts the signal from the mixing console so that it is loud enough for everyone in the room to hear.
- One or more speaker systems positioned around the room — usually at the front — to spread the amplified sound to everyone inside.

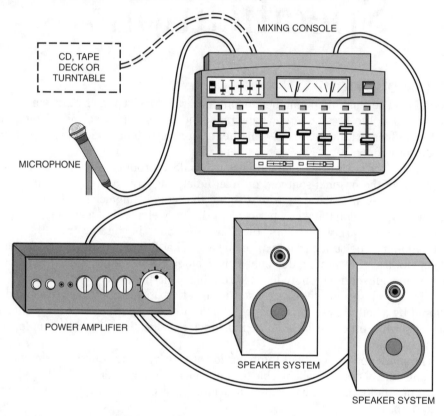

Figure 3-1. The four main components in a commonly used sound system are the microphone, mixing console, power amplifier, and speaker systems. (The mixing console may be omitted as was shown in Figure 1-1.)

"Hidden" components in all sound systems are the cables which connect everything together. Different kinds of cables are used for different applications. For example, one kind of cable is used to connect a microphone to the mixing console and another kind of cable is used to connect the power amplifier to the speaker systems.

Each of these major components — microphone, mixing console, power amplifier, and speaker systems — contributes to the overall quality of a sound system. The component specifications can help you assess the ability of each component to perform within the sound system's specifications. For instance, just by reading and understanding the specifications you can determine if a certain speaker system is appropriate or inappropriate for a given power amplifier.

MICROPHONE SPECIFICATIONS

Microphones are often the neglected component in the design of a sound system. While much attention is paid to the mixing console, power amplifier, and speaker systems, little attention is given to the microphones. This is decidedly a bad idea, because the proper choice of a microphone can make or break a sound system. A poorly selected microphone — also called "mike" for short — might produce a low and raspy sound. Unless you understand the technical specifications of the microphone, you might waste time checking other components to locate the source of a bad sound. Or worse, you might attempt to adjust for the poor performance of the microphone and end up over-compensating, making the sound even more unpleasant!

Microphone Element

The most important microphone specification is the kind of element in the microphone. The element converts sound pressure into electrical waves. The two most common microphone elements are dynamic and condenser.

Dynamic

The *dynamic* element, shown in *Figure 3-2,* is essentially a speaker driver working in reverse. Sound pressure moves a very thin diaphragm back and forth, which in turn, moves a coil of wire within the magnetic field of a permanent magnet. The movement induces an electrical voltage which produces a current in the coil circuit which changes in step with the sound pressure. The electrical signal is sent through the mixing console to the amplifier.

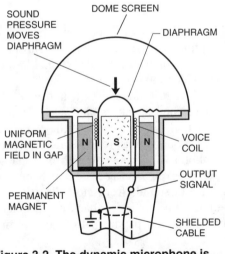

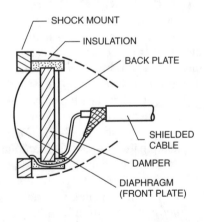

Figure 3-2. The dynamic microphone is basically a loudspeaker operating in reverse. Incoming sound pressure moves the diaphragm, which in turn moves a voice coil around a permanent magnet. The movement of the voice coil in the magnetic field induces a voltage in the voice coil proportional to the sound pressure.

Figure 3-3. The condenser microphone employs a thin diaphragm mounted on a metal plate. Air pressure causes the diaphragm to move in relation to a stationary metal plate, thus causing a change in capacitance.

Condenser

The *condenser* element, shown in *Figure 3-3,* is a more complex design and is often preferred in high-end sound systems. A diaphragm is attached to a metal plate. Behind this plate, but separated by an air gap, is another metal plate. Sound pressure moves the diaphragm back and forth, which changes the distance between the two metal plates. This creates a change in capacitance — hence the name "condenser" (condenser is another, older word for capacitor).

Condenser microphones are generally more sensitive than dynamic microphones, which means they can pick up softer sounds. A condenser microphone has a tiny preamplifier mounted close to the parallel plate capacitor. The preamplifier strengthens the electrical signal and matches the high impedance of the microphone element to the much lower impedance of the mixing console or amplifier. The preamplifier can be powered by the mixing console or amplifier to which it is connected, or more commonly, by a small battery in the microphone case.

Two other types of microphones that you may find in use are the electret and the piezoelectric microphone. Some systems favor the use of the *electret condenser* which is a special condenser microphone that has a built-in preamplifier. Piezoelectric microphones, which use a ceramic piezoelectric element, are actually a very old technology, and the only common use in modern sound systems is in a unique "conference room" microphone, as shown in *Figure 3-4.* This microphone picks up sound equally well in all directions, and delivers good sound quality even if the source is some distance from the microphone. Other types of microphone elements seldom found in new sound systems are crystal, carbon, and ribbon.

Sound Pickup Pattern

The second important microphone specification is its pickup pattern. The pickup pattern determines how sensitive the microphone is to sounds that come from all directions. There are two categories of pickup patterns: uni-directional, which means "one direction," and omni-directional, which means "all directions."

- *Uni-directional* microphones are most sensitive to sounds directly in front of the element. They are typically used for singers and speakers, and also for most instruments. They are often preferred in sound systems because they can help reduce acoustic feedback. *Figure 3-5* shows a graphic representation of the pickup pattern of a uni-directional microphone. A microphone with this characteristic heart-shape pick-up pattern is often referred to as a *cardioid* microphone.
- *Omni-directional* microphones are sensitive to sounds from most directions in the horizontal plane. It is not a good choice if acoustic feedback is a problem. *Figure 3-6* shows a graphic representation of the pickup pattern of a omni-directional microphone.

Bear in mind that the exact pickup pattern depends on the frequency of the sound. The pattern may change as the frequency increases. When a pickup pattern is graphically shown in the manufacturer's specifications sheet, it is usually for a "standard" test frequency of about 1 kHz.

Frequency Response

Recall from Chapter 2 that a sound system's frequency response is an indication of the sound system's ability to faithfully reproduce all the sound frequencies of

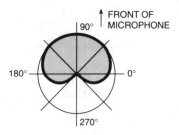

Figure 3-5. The pickup pattern of most uni-directional microphones is heart-shaped, hence the name "cardioid" microphone.

Figure 3-4. An omnidirectional piezoelectric microphone such as this model is often used in conference room setups where it is necessary to pick up the voices from many people situated around the room. (Courtesy of Radio Shack.)

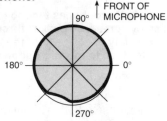

Figure 3-6. Omni-directional microphones pick up sound more or less equally in all directions.

human hearing — the range from 20 Hz to 20 kHz. In one sense, the microphone determines the overall frequency response of a sound system, because if the microphone cannot reproduce a given frequency, that frequency is missing from the sound as it travels through the remainder of the sound system.

As it turns out, microphones actually have the worst frequency response of all the components in a sound system. Take a look at *Figure 3-7,* which is a typical frequency response graph for a condenser microphone. It shows reasonable response starting at about 100 Hz, then dropping off at about 10 kHz. You should choose the microphone based on the source with which it is used. For voice, a mike with a frequency response of 100 Hz to 10 kHz should prove more than adequate, but for instruments, especially piano, a microphone with a broader frequency response is desirable.

You will also notice that the frequency response graph in *Figure 3-7* shows lots of hills and valleys. This is typical for microphones. The hills are more commonly called *presence peaks.* When selecting a microphone, look for a model that includes a graph of its frequency response. This will help you determine if a microphone is suitable for a given application.

Note from *Figure 3-7* that all microphones exhibit the *proximity effect.* This effect is heard as increased system bass (low frequencies) response as the source moves closer to the element. Singers prefer to get very close to the microphone, feeling it gives their voice a deeper, fuller sound. The proximity effect is seldom considered when the manufacturer measures and graphs the frequency response for a microphone; therefore, you will need to experiment with the microphone to determine if its proximity effect is acceptable for your application.

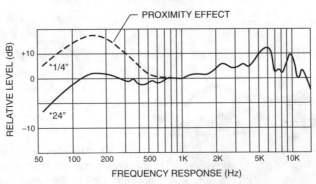

*Distance of Source from Microphone

Figure 3-7. Microphones tend to exhibit wildly fluctuating frequency responses, with many "presence peaks" throughout the frequency range. Also shown here is the "proximity effect," where the bass (low tones) are accentuated when the sound source is close to the microphone element.

Impedance

Impedance is one specification that has little to do with the overall quality of the microphone itself. Rather, impedance is important so that you correctly match the microphone to the proper input on the mixing console or amplifier.

Microphones have two general impedances: high and low. Most mikes for sound system and other professional applications have low impedance. Connect them to a low-impedance input on the sound mixer or amplifier. Be aware that the low-impedance match is not critical and that the impedance of a microphone changes with frequency.

High-impedance microphones tend to be the less expensive dynamic type found in low-cost tape recording applications. Be sure to connect the microphone to the high-impedance input of the sound mixer or amplifier; otherwise, the sound may be distorted. High-impedance microphones are much more likely to pick up noise from the microphone cable because of their high impedance.

Keep the following in mind when connecting a microphone to a mixing console or amplifier:

- Microphones with an impedance of 150 to 1000 ohms (600 ohms is common) connect to the low-impedance input. The actual impedance of this input is typically from 1k to 2k ohms.
- Microphones and guitar pickups with an impedance of 10k ohms or above connect to the high-impedance input. The actual impedance of this input is typically from 25k ohms to 250k ohms.

Other Microphone Specifications

Keep these additional specifications in mind when selecting the right microphone for the job.

- *Sensitivity.* The sensitivity of a microphone is typically expressed in dBm or dBV. The higher the dB value, the more sensitive the microphone.
- *Balanced/unbalanced connection.* Microphones with the familiar 1/4-inch phone plug or phone connector, as shown in *Figure 3-8a,* use an unbalanced connection. The connection is made using one wire and the shield as shown in *Figure*

3-8b — one for the signal and one for ground. Microphones with the three-prong "XLR" or "Cannon" connector, shown in *Figure 3-9a,* use a balanced connection, shown in *Figure 3-9b.* This type of connection, which helps reject electrical noise, is made with two wires and a shield — two for the signal and one for ground.

- *Transient response.* The ability of a microphone to respond to sudden changes in sound — like a sudden, loud clap — is known as transient response. Transient response is much more important for musical reproduction than for voice reproduction.

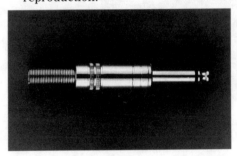

a. Phone Plug

a. XLR Connector

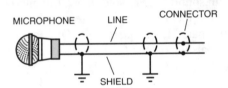

b. Schematic of Unbalanced Circuit   b. Schematic of Balanced Circuit

Figure 3-8. The 1/4-inch phone plug is commonly used on low- and mid-priced microphones. It employs a pair of wires: signal and ground. It is used in unbalanced input connections.

Figure 3-9. The three-conductor "XLR" connector is often used with higher-priced microphones. It employs three wires: two signal and one ground. It is used in balanced input connections.

MIXING CONSOLE SPECIFICATIONS

A mixing console, such as the model shown in *Figure 3-10,* is used to combine and control the level of two or more inputs, such as several microphones, a cassette player, a CD player, and a broadcast radio tuner. The console lets you adjust the sound level of each source to tailor the output level of the combined signals as desired. For example, you can reduce the level of one source and increase the level of another source to achieve an overall effect that is more natural and pleasing to the audience.

A mixing console usually has a preamplifier at its input to boost the signal level just enough to keep the signal-to-noise ratio high while it performs the mixing function. A power amplifier is then used to amplify the relatively low-level signal from the mixing console for delivery to the speaker systems. Some mixing consoles also incorporate a power amplifier; however, in this book, we will consider the mixing and amplifying functions as separate.

Figure 3-10. Mixing consoles tend to look more complex than they really are. This model accepts up to six inputs, including two microphones. Slide faders adjust the sound levels.

Type and Number of Inputs

Mixing consoles differ in the number of inputs they provide. The ones designed for the typical small sound system usually provide two inputs for microphones, and three to five inputs for line-level sources such as CD players, tape decks, turntables, or broadcast radio tuners. In almost all sound mixers sold today, the microphone inputs are monaural and the line-level inputs are stereo. If stereo, you can usually control the level of the right and left channels independently. One mixing console can be coupled through line-level signals to another to expand the number of particular inputs.

Number of Outputs

Most mixing consoles for small- to medium-size sound systems provide one output (the console may be capable of stereo, with separate left and right channels, but it is still considered to have just one output). This single output is connected to a power amplifier.

Some high-end mixing consoles provide more than one output. The additional outputs have various uses, but a typical use is piping the sound, at a reduced level, through monitor speakers situated where the performers are located. These speakers face the performers, and are designed to give the performers a better notion how the audience is hearing their performance.

Level Controls

The primary goal of a mixing console is to mix sources to blend them into a harmonious sound. The mixing is done by controlling the levels of the individual sources. The level controls are called *faders*, because they are used to boost or fade the level of an input. A popular type of fader is the slide control (or "slide pot") as indicated in *Figure 3-11*. You operate the control by sliding it back and forth. The slide action is easier than turning a knob, and you will grow to appreciate it when controlling the sound levels during a concert or other presentation.

Each input has its own fader, which allows you to vary the level from zero (no sound) level to full level. For stereo mixing consoles, each of the line-level inputs

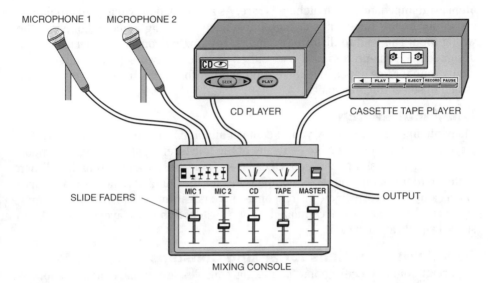

Figure 3-11. The faders of a mixing console permit independent control of the relative levels of each input, such as microphones, CD player, and tape deck. The master fader varies the output level of all sound sources simultaneously.

have two faders: one for the right channel and one for the left channel. In addition, the mixing console provides a master level control, called the *master fader*. When you vary the level of the master fader, it changes the output level of the combined sources.

Additional Controls

Depending on the mixing console, it may contain additional controls. The most common ones are:

- *Equalizer.* Equalization (or EQ) controls let you tailor the sound of particular frequencies being output to the speakers. The equalizer controls let you increase or decrease the level of the sound within discrete frequency ranges. For example, if you think the amplified sound is too boomy, you can use the equalizer to decrease the level of the low frequencies and increase the level of the high frequencies.
- *Echo and delay.* Echo adds reverberation to the sound, and is typically used to make the sound seem more full. It's most often used for special effects, but some vocal groups like it as well. Delay, if offered, delays the sound a bit, and can create the illusion of a larger room.
- *Pan.* Panning lets you direct an input, like a microphone, to either the left or right speakers, in varying degrees. You can use the pan (or commonly called *balance*) control, for example, to send the voice of one singer to the left speaker, and the voice of another singer to the right speaker.

Input Impedance

The inputs of the mixing consoles are designed to match (at least in general) the impedance of the input device connected to it. Line-level inputs are usually 600 ohms, and impedance mismatching is rare. As mentioned previously, microphones can either be low impedance or high impedance. Coupling a microphone into the wrong input can lead to distorted sound. Many mixing consoles, especially those designed for semi-professional use, either provide for two separate microphone inputs (one for low-impedance mikes and another for high-impedance mikes), or a switch to select the impedance.

Line-Input Switch

Many mixing consoles allow you to connect any line-level device — such as a CD player, tape deck, or turntable — to any of the line-level inputs. Of greatest importance is distinguishing between CD players/tape decks and turntables. When switched to turntable, the mixer compensates for the characteristic response curve of a pick-up head cartridge (this response curve is known as *R.I.A.A. equalization*), and low frequencies are accentuated. When switched to CD player/tape deck, no special equalization is used.

Sound Specifications for Mixing Consoles

Important sound specifications for mixing consoles are signal-to-noise ratio, frequency response, dynamic range, and headroom. These sound specifications were discussed in Chapter 2.

AMPLIFIER SPECIFICATIONS

Power amplifiers can be either stand-alone, such as the model shown in *Figure 3-12,* or, as mentioned previously, they can be combined with a mixing console. Whether separate or combined, the same specifications and considerations apply.

Power Output

Power output is rated in watts. The higher the watts, the more power the amplifier is capable of delivering; therefore, the louder the sound (given speaker systems that can handle the power as discussed later in this chapter). Don't be too quick to judge an amplifier on power output alone. It's very important to match the power output of amplifier with the amplification needs of a particular environment, whether it be a small conference room or a large high-school gymnasium.

Figure 3-12. A power amplifier is the cornerstone of the sound system. Here is a stereo model that provides 100 watts of RMS power to each of the right and left channels. (Courtesy of Radio Shack.)

Each case is separate and the unique acoustics of the room plays an important role in the power output requirements of an amplifier. A room that tends to be "dead" to sound (little reverberation) will require an amplifier with more output power. In a "live" room, amplifier with a 25- or 35-watt output can deliver an amazing amount of sound, and it may be more than enough for your application. Bear in mind that with higher wattage comes a higher price. There is no sense in paying for wattage you will never use.

Table 3-1 provides some basic guidelines for power amplifier wattage. In this table, the following definitions apply:

- A small room is about 1,500 square feet which can accommodate up to 100 people. This is about the size of a corporate conference room.
- A medium room is about 1,500-4,500 square feet which can accommodate up to 300 people. This is about the size of a lodge or community meeting hall.
- A large room is over 4,500 square feet which can accommodate over 300 people. A typical large room is a high-school gymnasium or a large assembly hall.

Table 3-1. Power Required for Various Room Sizes

Room Size	Total Power Amplifier Output
Small	20 watts
Medium	35 watts
Large	100 watts

Some sound system installations require a closer match of power to room size than given in *Table 3-1*. Using the proper math, it is possible to determine more accurately the power output requirements for a given room by considering the size, shape, and acoustics of the room; the size, number, and type of loudspeakers used; and the typical program material amplified by the sound system. However, these calculations are beyond the scope of this book for your specialized sound system. Seek the help of competent sound systems professionals; they can help you choose the proper components.

Power Output: RMS or Peak?

It is customary for makers of power amplifiers to list the power output in watts RMS. RMS stands for *root mean square*. It is a more complex approach than simple averaging and yields more accurate results. Power outputs of other forms are listed for some amplifiers.

Peak (sometimes referred to as "maximum") output is the highest wattage the amplifier can deliver in a short burst of time. It is not unusual for an amplifier to be capable of delivering a peak output of twice its rated RMS wattage. The peak power capability helps to handle the transient peaks of sound we talked about earlier.

In the power rating for an amplifier, you will typically see a notation such as "100 watts into 4 ohms." This means the power of the amplifier was tested with a 4-ohm load. The "ohms" notation is important because it helps you compare amplifiers. An amplifier capable of delivering 100 watts into 4 ohms is the same as an amplifier capable of delivering 50 watts into 8 ohms (or 25 watts into 16 ohms).

Eight-ohm loudspeakers are the most common, but 4-ohm and 16-ohm models are also available.

Output Terminals and Impedance

All power amplifiers have at least one set of output terminals to connect to a speaker system. The output terminals are usually physically large to accommodate the high-power output.

Many power amplifiers have multiple output terminals, as shown in *Figure 3-13,* giving you the freedom of connecting to speaker systems of different impedances. Typical speaker system impedance is 8 ohms, however, some speaker systems are rated at 4 ohms or 16 ohms. You must connect your speaker system to the proper output terminal or damage to the amplifier or speaker system may result. At any one time, you can only connect your speaker systems between *one* of the output terminals and ground.

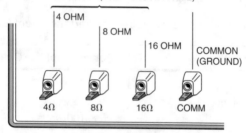

Figure 3-13. Many power amplifiers provide separate terminals for connecting to 4-, 8-, and 16-ohm speaker systems. A single common ground terminal is used for all three impedances. Only one terminal and ground can be used at a time.

It is possible to connect two or more speaker systems to a single output terminal. When combining speaker systems, the impedance presented to the amplifier is either halved or doubled, depending on how they are connected (this topic is covered in Chapter 5). You need to calculate the effective speaker system impedance and connect the speaker systems to the proper terminal. Again, failure to do so may lead to damage to the amplifier, speaker systems, or both.

Some sound system amplifiers also provide a set of terminals for connecting to a 70-volt constant-voltage distribution system. This system is typically used for paging and background music, where high sound volume is not the main consideration. Details of this system are given in Chapter 5.

Sound Specifications for Power Amplifiers

As with mixing consoles, common sound specifications for power amplifiers include dynamic range, frequency response, and signal-to-noise ratio. These sound specifications were discussed in Chapter 2. In addition, the following specifications are provided for many power amplifiers.

Total Harmonic Distortion

Total harmonic distortion (THD) measures the ability of an amplifier to reproduce a signal without contributing additional frequency components called *harmonics.*

Harmonics are frequencies that are mathematically related to the *fundamental* frequency.

Total harmonic distortion is measured as a percentage of the newly created harmonic frequencies in relation to the original signal. The smaller the percentage, the better the amplifier. The measurement varies as both the level (amplitude) of the sound and its frequency varies, so tests are made with specific tones and certain predetermined levels. A typical test specification of THD for a sound system power amplifier might read "0.05% at 1 kHz," meaning there is 0.05% or less THD when tested at 1 kHz.

Intermodulation Distortion

Intermodulation distortion (IM) is produced when a sound circuit creates new frequencies by mixing those being processed. This can cause unfavorable side effects, making the sound "off timbre."

If two frequencies are input to the amplifier, a series of sum and difference frequencies will be generated by a circuit producing IM. For example, if a 100 Hz signal and a 1000 Hz signal are input to the amplifier, the amplifier generates a difference signal of 900 Hz and a sum signal of 1100 Hz. In reality, many other less intense frequencies are created out of these new ones, and the effect goes on for as long as the amplifier circuitry allows.

Like THD, IM is expressed as a percentage. The smaller the percentage, the better the amplifier. IM distortion is more objectionable than THD, but an excellent power amplifier exhibits low percentages of each.

Crosstalk

Stereo amplifiers must boost the right channel and the left channel signals independently. Virtually no amplifier can process both channels without some signal from one channel getting into the signal from the other. When this happens, the amplifier is said to exhibit *crosstalk* (also called *channel separation*).

A typical crosstalk specification might be –90 dB. The –90 dB means the signal from the opposite channel is 90 dB *less* in intensity than the signals from the present channel. If a frequency is not specified, it is usually safe to assume the test was made at either 1 kHz or is an average of tests at various frequencies.

LOUDSPEAKER SPECIFICATIONS

It all comes together at the loudspeakers. The loudspeakers of a sound system convert the amplified signals to sound. By their nature, loudspeakers tend to be simple devices, yet the science of turning electrical impulses to music is a complex one. We will touch only on the most critical specifications for loudspeakers in this section. A cross-section of a typical loudspeaker is shown in *Figure 3-14*.

Size

The size specification for loudspeakers refers to the cone diameter. Loudspeakers come in many different sizes, from half-inch miniature models designed for headphones to leviathans of more than two feet. However, for the usual sound system found in schools, churches, and offices, most loudspeakers are between 8 and 15 inches.

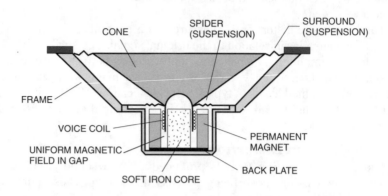

Figure 3-14. Loudspeakers work by applying a signal to a voice coil, which surrounds a permanent magnet. The moving voice coil vibrates the cone to produce sound.

Loudspeakers work by vibrating a cone which, in turn, vibrates the air in front of and behind the cone to produce sound. Generally, the larger the cone, the larger volume of air it moves, hence, the louder the sound. High frequencies don't need large loudspeakers to produce loud sounds, but low frequencies do. The lower the frequency, the more energy is needed to move the air; therefore, physically large loudspeakers are required. Loudspeakers are placed in enclosures to improve the speaker system's frequency response, efficiency and sound output — especially at low frequencies.

In many sound system installations, the speaker systems are designed to reproduce all of the sound frequency range. Sometimes only one fairly large loud-speaker does all the work. Another popular design is a speaker system with two loudspeakers. One was shown in *Figure 1-1*. It has one large loudspeaker for the combined low and mid-range frequencies and another loudspeaker for the high frequencies. The highs are reproduced by a tweeter, and is usually in the shape of a horn. The horn shape helps disperse the sound.

There is a practical limit to the size of a loudspeaker. Though it is possible to design and manufacture loudspeakers over two feet in diameter, a more economical and practical approach is to simply use additional loudspeakers. Two 15-inch loudspeakers move about the same amount of air as a single 30-inch loudspeaker. In sound systems for large meeting halls or gymnasiums, it's not unusual to find speaker systems with an array of loudspeakers — perhaps four or five 12- or 15-inch loudspeakers in a single enclosure.

When choosing speaker systems for a given room, you need to consider both the size and quantity of the loudspeakers inside the enclosures. *Table 3-2* gives some basic guidelines. The definitions for small, medium, and large are the same as for *Table 3-1*. The "Room A" configuration is a room with a width-to-length ratio less than 1.5:1 (almost square). The "Room B" configuration is a room with a width-to-length ratio between 1.5:1 and 3:1. Both configurations assume a ceiling height of less than 20 feet.

Table 3-2. Loudspeaker Quantity and Size for Various Room Sizes

Room Size	Room A		Room B	
	Qty.	Size	Qty.	Size
Small	2	8"	4	8"
Medium	2	12"	4	12"
Large	4	12"	6	12"

Power Rating

Loudspeakers convert electrical energy into sound. In doing so, they create heat. This heat is dissipated from the loudspeaker into the air or conducted to a special heat-conductive fluid surrounding the loudspeaker voice coil. Too much heat will burn out a loudspeaker, so it is important to monitor the power delivered by the amplifier.

All loudspeakers are rated by their power handling capacity in watts. A typical sound system loudspeaker might be rated at 100 watts, meaning that the loudspeaker can handle 100 watts of power before there is a danger of burn out.

There are actually two kinds of power ratings — RMS and peak — as we discussed earlier in this chapter. RMS denotes the wattage the loudspeaker can handle on a continuous basis. Peak power denotes the wattage the loudspeaker can withstand on an intermittent basis. It's not uncommon for a loudspeaker to handle twice the peak power as RMS power.

As we have stated several times previously, it is important to match the speaker systems to the power amplifier. As a rule of thumb, always select loudspeakers that are rated *at least as high* as the power output of the amplifier. For example, if the amplifier is capable of delivering 80 watts RMS, select loudspeakers that are rated at no less than 80 watts RMS. Failure to do so could lead to blown out loudspeakers!

Impedance

All loudspeakers are rated by their impedance. The majority of loudspeakers are rated at 8 ohms; however, there are a number of specialty loudspeakers rated at 4 ohms, and a few at 16 ohms.

The impedance of the loudspeakers is important because it determines the amount of power delivered to them by the amplifier. An amplifier capable of delivering 100 watts into an 8-ohm loudspeaker will actually provide approximately 200 watts of power into a 4-ohm loudspeaker. If the 4-ohm loudspeaker were rated for 100 watts, it could be damaged. We showed in *Figure 3-13* how many sound system amplifiers provide multiple terminals for connection to 4-, 8-, or 16-ohm impedance speakers.

Sound Specifications

The sound a loudspeaker reproduces is influenced by its dynamic range and frequency response specifications, which were discussed in Chapter 2. Other loudspeaker specifications are briefly explained below.

- *Free-air resonance* is the frequency at which the sound-producing material of the loudspeaker (i.e., the cone) vibrates the easiest. This resonance is determined by the mass of the cone, the size of the cone, and other elements.

- *Qts* denotes the resonance magnification of a loudspeaker, which is the tendency of a loudspeaker to reach its maximum sound output level when operating at the free-air resonance frequency.
- *Vas* measures how compliant the loudspeaker cone is, and therefore, how readily it moves air in response to electrical signals. A "stiff" cone has a low compliance; a "loose" cone has a high compliance.
- *SPL,* or sound pressure level, indicates the volume of sound produced by the loudspeaker. The higher the SPL, the louder sound the loudspeaker can generate from a given signal.

There are other specifications and design criteria for loudspeakers, but those discussed above should be sufficient for analyzing your sound system needs. If you would like to learn more about loudspeakers and speaker system design, refer to *Speakers for Your Home and Automobile* and *Advanced Speaker Designs*, both published by PROMPT Publications, Inc. *Speakers for Your Home and Automobile* is intended for home systems; *Advanced Speaker Designs* offers more detailed speaker systems design.

UPGRADE OR BUY ALL NEW EQUIPMENT?

It is much easier to purchase the components for an all new sound system than it is to upgrade one or two components in an existing sound system. When you buy all new equipment, it is easier to match the components. For example, you can ensure that the speaker system and the power amplifier are rated properly, and that the mixing console has enough inputs for your application.

Buying new gear isn't always an option because of the high cost. Rather than buy everything over again, you are faced with upgrading just those components that need replacing. Perhaps the power amplifier is no longer functioning and cannot be repaired at a reasonable cost. Or perhaps the loudspeakers are damaged or worn from years of use.

Whatever the reason, you must exercise care to choose a replacement that blends in with the rest of the system. If you are replacing the power amplifier, for instance, there is no sense in buying an expensive 200-watt model if your speaker system is rated at only 50 watts.

Consider also the quality of the existing components. There is little justification in purchasing top-of-the-line speakers only to install them on a bargain sound system that is equipped with a single microphone and a 25-watt amplifier. Conversely, don't stymie yourself by under-buying. If your sound system requires a new mixing console with five or six inputs, don't settle on a less-expensive and less-capable model just to save a dollar here or there.

SUMMARY

This chapter reviews the specifications of the four major parts of a sound system: microphones, mixing consoles, power amplifiers, and speaker system. You learned how to understand the most commonly used specification. By knowing more about specifications, you know how to gauge the way each individual component affects the sound system. This will help you choose new equipment and maintain an existing system.

Installing Sound System Components

INTRODUCTION

Most public sound systems are permanent installations; therefore, it's important to think seriously about the arrangement of the system components. A haphazard component arrangement can degrade the performance of even the most expensive and feature-laden sound system.

Different component arrangements are needed for rooms of different shapes and sizes. In fact, the room itself may greatly determine how and where the components are installed. In a small room, for example, sound system components can often be placed in a centralized location. For larger rooms, it is usually necessary to separate at least part of the equipment. This raises such factors as cable and speaker wire lengths, and the placement of microphones and speaker systems, especially to prevent acoustic feedback.

This chapter discusses the installation of the sound system's main components: mixing console, power amplifier, and speaker systems. We will review common and accepted strategies for placement of these components in rooms of various shapes and sizes. Though it is impossible to detail the perfect sound system setup for all kinds of rooms, sound system dynamics for the types of systems covered by this book are not rigid. With some forethought, you should be able to adapt one of the described room categories to your unique situation.

In this chapter, we will concentrate only on placement and installation of the main components of the sound system. The details of connecting the electrical wiring of these components together will be discussed in the next chapter.

SOUND SYSTEM COMPONENT GROUPS

So far in this book, we have categorized sound system components in three broad groups: input, signal processing, and output. For system installation, we will use these same three groups, but, as shown in *Figure 4-1,* change the mix of the components that fit into the groups.

- The input group comprises the sound sources: microphones, turntable, compact disc player, and audio cassette player.
- The signal processing group comprises the mixing console and equalizer. The equalizer may be part of the mixing console or a separate component. Recall that the mixing console blends the sounds from the various signal sources, while the equalizer changes the overall signal response within various frequency bands.

- The output group comprises the power amplifier and the speaker systems. This portion of the system delivers the proper sound levels to the assigned destinations.

The power amplifier and speaker systems are combined into the output group because these two components need to be close together. A great deal of power can be lost if they are not. This situation is discussed in this chapter.

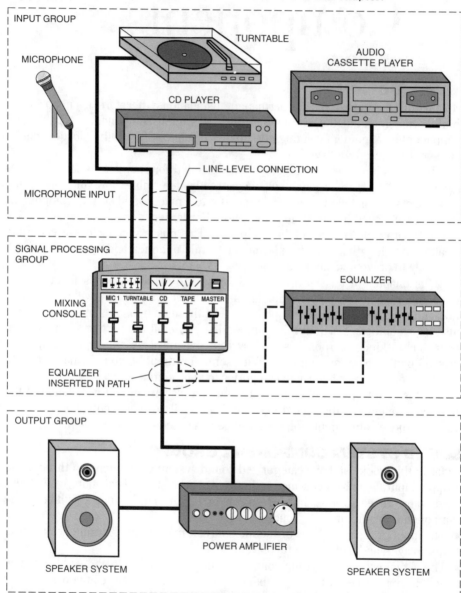

Figure 4-1. For installation purposes, you can consider the components of a sound system separated into three groups: input group, signal processing group, output group.

ROOM SIZE DEFINITIONS

For the sound system installations covered in this book, rooms — conference rooms, meeting halls, churches, and auditoriums — are divided into three general size categories: small, medium, and large. Our definitions — approximate, of course — of "small," "medium," and "large" room sizes are shown in *Table 4-1.*

Table 4-1. Definition of Room Size

Room Size	Square Feet	Approximate Audience Size
Small	Up to 1,500	Up to 100 people
Medium	1,500 to 4,500	Up to 300 people
Large	Over 4,500	Over 300 people

Shape of the room will definitely influence the placement of the speaker systems, and maybe the amplifiers, but generally not the other components.

Very large rooms — greater than 15,000 square feet — often require specialized sound system installations. The design and application of such sound systems is beyond the scope of this book. If you are in charge of a sound system for a very large room, you may want to consult with a sound system professional for guidance in selecting and installing the equipment.

Special consideration sometimes is needed if the room, whatever its size, incorporates a thrust stage. A thrust stage is a stage that extends into the audience area of the room. We will consider rooms with stages as a separate sub-group.

AMPLIFIER-TO-SPEAKER DISTANCE

As we review the overall design approaches for the three room sizes, a central issue will become clear — minimizing the distance between the amplifier and speaker systems. Why is this so important?

Amplifiers and speaker systems are connected by wires. Current in the wires delivers energy from the amplifier to the speaker systems. Recall that all wire has resistance and that resistance opposes the flow of electrical energy. For a given material, wire resistance is primarily dependent on the length and diameter of the wire. Resistance increase as the wire length increases or as the wire diameter decreases. To emphasize the importance of the distance between amplifier and speaker system (the length of the connecting wires), consider the general rule of thumb is that resistance doubles for each doubling of wire length.

The resistance of the wire dissipates electrical power in the form of heat; therefore, it reduces the effective power delivered to the speaker system. For example, if an amplifier puts out 40 watts, but the speaker wire resistance dissipates about one-third of that power, then only about 27 watts are delivered to the speaker system.

An equally important factor is the wire diameter. The larger the diameter of the metal wire (not including the insulation), the smaller its resistance. Think of the wire as a water pipe; the larger the pipe diameter, the more water can flow through the pipe in a given time. Similarly, the larger the wire diameter, the more electrons can flow through the wire in a given time; therefore, the less resistance the wire exhibits.

4

All wire, including speaker wire, is rated by gauge. The smaller the gauge; the larger the wire. Always choose the smallest gauge (largest wire) practical, because it will have the least power loss. *For sound systems, don't use wire greater than 16 gauge for connecting the power amplifier to the speaker systems.* This is adequate for a 25- to 35-watt sound system in a small room where the wire length is *under 100 feet.* Smaller gauge wire (larger wire, smaller resistance) is required for higher wattage and/or longer runs, as indicated in *Table 4-2.*

Table 4-2. Recommended Wire Gauge for Speaker Systems

Maximum Wattage	Minimum Gauge (for 100-foot wire length maximum)		
	4-ohm Speaker	8-ohm Speaker	16-ohm Speaker
25	14	16	16
50	12	14	16
100	10	12	14

Speaker wire larger than 12 gauge is available, but is harder to find and is expensive. It is generally more economical and more practical to reduce the wire length. Of course, this requires that the power amplifier be closer to the speaker systems. One other point of note: speaker wire normally has a stranded conductor (comprised of several small diameter wires) to make it flexible and easy to install. Solid conductor wire, especially of smaller gauge, is stiff and harder to install.

SOUND SYSTEMS FOR SMALL ROOMS

Sound systems for small rooms can be more difficult to design and install because there is less room — literally — for error. The most annoying problem is acoustic feedback because it is difficult to physically separate the microphone a great distance from the speaker system.

The shape of the room can be a consideration in placing the mixing console (if used) and the power amplifier. The small size of a square room makes it impractical to position the console and power amplifier near the front, especially if you (or someone else) must monitor the program and control the mixing console. *Figure 4-2* shows an ideal arrangement of the sound system components. Note that the signal processing equipment and the power amplifier are located near the back of the room. The microphone is on a podium in the front-center of the room. The speaker systems are placed in front of the microphone to avoid acoustical feedback. The microphone cable to the mixing console is a shielded cable. This arrangement is preferred so that it is less intrusive to the audience.

For a small room, the distance between power amplifier and speaker is not critical (not to mention the typical lower output wattage). Even in the worst case — with a speaker in one corner and the power amplifier in the other corner — the distance is only about 45 feet. If the recommended wire gauge given in *Table 4-1* is used, this distance should not significantly reduce the speaker system's output. The speaker wire connections are shown in *Figure 4-2,* but in subsequent room layouts, the speaker wire to amplifier connections are not shown.

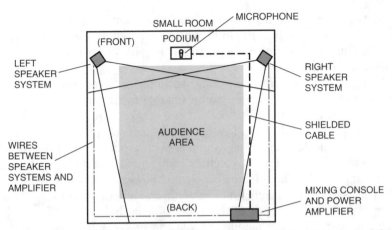

Figure 4-2. In the basic small room installation, both the mixing console and the power amplifier can be located in the back of the room, out of the way of the audience. The speaker systems are placed so that they point to near the center of the audience area.

SOUND SYSTEMS FOR MEDIUM ROOMS

As the room size increases, the distance between the power amplifier and speaker systems may also increase. Since speaker systems generally must be located in (or near) the front of the room, this somewhat limits the practical location of the power amplifier. *Figure 4-3* shows a medium-sized rectangular-shaped room. The mixing console and power amplifier are placed on one side of the room near the wall. This

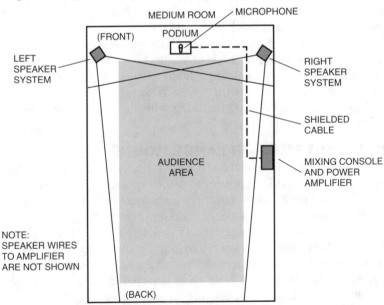

Figure 4-3. In a medium room, it may be necessary to position the mixing console off to one side of the room to reduce the amplifier-to-speaker distance. It is considered good practice to limit the amplifier-to-speaker distance to no more than 100 feet.

placement reduces the intrusion of the equipment (and operator) on the audience. Again, the microphone is at the front-center on a podium and the speaker systems are placed in front of the microphone. The cable from microphone to mixing console is a shielded cable.

With this arrangement, the worst-case amplifier-to-speaker distance is usually under 75 feet. If you use a wire gauge suitable for the amplifier's power output, there should be minimal loss in speaker wire of this length.

Side placement of the sound system mixing console and amplifier is not always possible, or even desirable. For example, such a placement is not suitable for a theater auditorium, where all electrical equipment should be hidden from the audience if possible. The mixing console and power amplifier are typically located at the back of the auditorium and usually inside an enclosed room (or possibly a service closet). If the depth of the room is less than about 75 feet, it may be possible to place both the mixing console and power amplifier at the back of the room.

On the other hand, if the room is more than 75 feet deep, the amplifier-to-speaker distance may be in excess of 100 feet (remember that extra wire length is needed for turning corners, for tucking away in basements or ceilings, and for slack). In such a situation, there are these possibilities:

1. You can use a smaller gauge speaker wire, but if this is not possible or economical, then;
2. Use the preferred way of positioning the power amplifier close to the speaker systems as shown in *Figure 4-4*. This increases the length of the wire connection between the amplifier and the mixing console (which is a low-energy connection), but reduces the length of the more critical high-energy connection between the amplifier and speaker systems.

When the power amplifier is moved more than a few feet from the mixing console, be sure to use fully shielded connections (as used for the microphone cable) between them. The shielding helps prevent static and other noise from entering the system by way of the wires. With proper shielding, the connecting cables can be as much as 100 feet long with no appreciable loss of power or degradation of the signal.

SOUND SYSTEMS FOR LARGE ROOMS

The typical large room is an auditorium, church, or meeting hall. For these rooms, it is usually not desirable to have the sound system equipment exposed to the audience. Therefore, a "sound room" of some type is almost always employed to house the sound system components. (Lighting control equipment, if any, may also be located in this room.) The sound room is situated at the back of the auditorium.

Because the distance between the sound room and the speaker systems is almost always more than 100 feet, the power amplifier should be placed near the front and close to the speaker systems as depicted in *Figure 4-4*. If this simply is not feasible, then a larger speaker wire (smaller gauge wire) than that shown in *Table 4-2* must be used. In this situation, amplifier power will be lost in the wires connecting the power amplifier to the speaker systems; therefore, a power amplifier with higher output power may be necessary to deliver the proper sound output to the speaker systems.

4

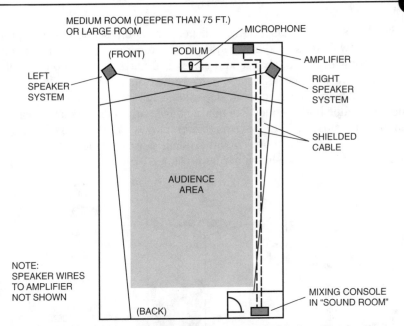

Figure 4-4. In large rooms (and medium rooms where a side location for the sound system components is not possible), the amplifier can be placed close to the speaker systems, and the mixing console — as well as any other input component like a turntable or CD player — can be placed in the back of the room.

SPEAKER DISPERSION

What is dispersion? Consider the speaker systems shown in *Figure 4-5*. Speaker system A has a wide dispersion so the sound spreads out more as it leaves the speaker. Because the audience in a wide, shallow room is likely to be very close to the speaker systems, using wide dispersion speaker systems helps to spread the sound before it reaches the audience.

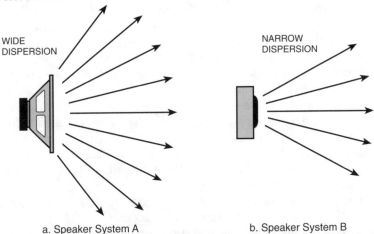

Figure 4-5. All speaker systems exhibit dispersion. Wide dispersion speaker systems spread sound over a larger area. Narrow dispersion speaker systems (the most common for sound system installation) limit the spread of sound. By limiting the spread of sound, it can travel further.

Speaker system B has a narrow dispersion so the sound doesn't spread very much. Narrow dispersion speaker systems are recommended for most sound systems in rooms of typical shapes and sizes. The narrow dispersion permits better directional control of the sound and helps reduce acoustic feedback.

Sound dispersion is frequency dependent. Low frequencies tend to have wide dispersion, while high frequencies tend to have narrow dispersion. Because of this phenomenon, most speaker systems that employ separate drivers for the different frequency ranges use a "horn" or other such device for better dispersion of the high-frequency sound as shown in *Figure 4-6*.

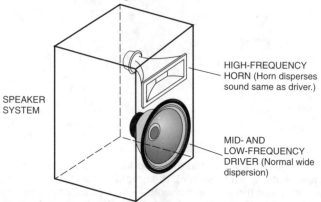

HIGH-FREQUENCY HORN (Horn disperses sound same as driver.)

SPEAKER SYSTEM

MID- AND LOW-FREQUENCY DRIVER (Normal wide dispersion)

Figure 4-6. This speaker system has a high-frequency horn and a mid/low-frequency driver. The horn helps disperse the high frequencies, so that the higher frequencies spread out as much as the lower frequencies from the mid/low-frequency driver.

PLACING THE SPEAKER SYSTEMS
Placing Speaker Systems in the Front Corners of the Room

In the room illustrations for this chapter, we've shown speaker systems situated near the front corners of the room. This is a common design. Two speaker systems are used, even though the sound system may not be stereo, because this arrangement disperses the sound over the entire audience area. The proximity of the speaker systems to the corner walls helps increase the sound output.

Installing the speaker systems in the front corners of the room may not always be possible, for the following reasons:

- A thrust stage juts out into the audience area as indicated in *Figure 4-7*. If a microphone is placed toward the front of the stage and is located in front of the speaker systems, acoustic feedback will most likely be a problem. As discussed previously, probably the simplest way to reduce or eliminate acoustic feedback is to move the speaker systems forward as indicated in *Figure 4-7* (this works if the microphone is a directional type).

- The room has hard walls, such as brick or concrete. Placing the speaker systems too close to the corners may cause excessive sound reflections, which under some circumstances can make the program less intelligible. In such rooms, it is usually necessary to increase the distance between the speaker system and wall, or add acoustic damping material (a quilt or other heavy cloth) on the wall near the speaker system as shown in *Figure 4-8*.

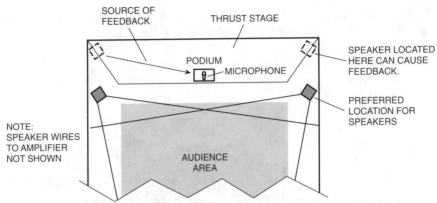

Figure 4-7. In room with thrust stages, avoid placing the speaker systems in the very front corners. Otherwise, amplified sound from the speaker systems could re-enter a microphone placed on the stage and cause acoustic feedback.

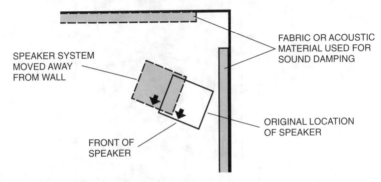

Figure 4-8. To reduce unwanted reverberation, it may be necessary to add more space between the speaker system and the wall(s). You can also add fabric for sound damping.

Placing Speaker Systems in Rooms with Unusual Shapes

The typical room for a sound system has a rectangular shape with the depth being slightly greater than its width. The depth to width relationship is expressed as a ratio. For example, if the ratio is 1.25:1, the depth of the room is 1.25 times its width; so, if a room is 20 feet wide, the depth is 1.25 times 20, or 25 feet.

Rooms with a depth to width ratio of between 1:1 (square) and 1.5:1 are most receptive to a basic sound system installation. A room with greater than 1.5:1 often requires special treatment. So, too, does a room that has a depth to width ratio of less than 1:1, which means that it is wider than it is deep. Fortunately, these kinds of rooms are rare.

Depth to Width >1.5:1

Rooms with a depth to width ratio exceeding 1.5:1 often require higher output amplifiers and additional speaker systems to ensure adequate coverage. The general idea is to present approximately the same sound level to the audience in the back as to the audience in the front. *Figure 4-9* shows a room with a depth to width ratio of 2:1. Rather than crank up the volume of the speaker systems located in the front of the room — which would blast the first few rows of the audience — additional speaker systems are placed along the sides of the room. These speaker systems are typically located a little less than half way along the depth of the room. (We're assuming a room less than 150 feet long, where the sound delay — due to the relatively slow speed of sound — between the front and side speaker systems is not noticeable.)

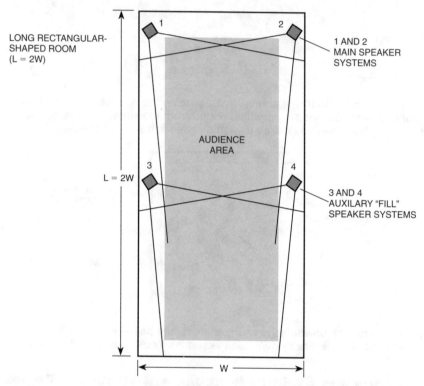

LONG RECTANGULAR-
SHAPED ROOM
(L = 2W)

1 AND 2
MAIN SPEAKER
SYSTEMS

AUDIENCE
AREA

L = 2W

3 AND 4
AUXILARY "FILL"
SPEAKER SYSTEMS

W

Figure 4-9. In rooms that are very long with respect to their width, it is often necessary to add auxiliary or "fill" speaker systems along the length of the room. Such an arrangement provides better coverage and helps prevent sound level imbalances (too high in the front, too low in the back) throughout the audience area.

Most often, the speaker systems along the side of the room are physically smaller than the main speaker systems and produce a lower level sound to supplement the sound from the main speaker systems. (This arrangement is similar to the use of a fill light in photography to reduce or blend the shadows from a key light.)

Depth to Width >3:1
If the room is large and exceptionally deep (up to about a 3:1 depth to width ratio), even more speaker systems may be required to distribute the sound at about the same level to all areas of the audience. The additional speaker systems are often situated along the side of the room and spaced for the best distribution of sound.

Depth to Width <1

Special placement of speaker systems is also required for rooms that are wider than deep (depth to width ratio is less than 1). Some lecture halls and classrooms are designed this way to reduce the maximum distance between instructor and student. The usual arrangement of two speaker systems in the front of the room is often inadequate for this shape of room. Instead, additional speaker systems with wide dispersion are placed along the width of the front of the room as shown in *Figure 4-10.*

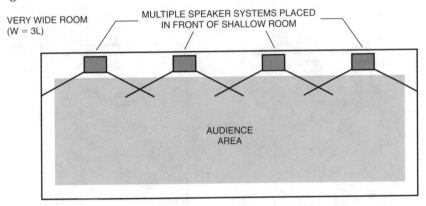

Figure 4-10. In rooms that are wider than they are deep, it is necessary to install multiple speaker systems along the front to adequately cover the entire audience area.

Do's and Don'ts of Speaker Placement

The overall performance of a sound system depends a great deal on the placement of the speaker systems. Haphazard speaker placement can handicap an otherwise excellent sound system. Keep the following points in mind when placing speaker systems for best sound coverage; in all cases, at least two speaker systems are assumed.

- *Do* aim the speaker systems over the central portion of the audience. This usually involves turning the speaker systems in at the front corners of the room as illustrated in *Figure 4-11a.*
- *Don't* mount the speaker systems so that the sound output is parallel to any wall, as shown in *Figure 4-11b.* Excessive reverberation could result.
- *Do* use additional speaker systems along the depth of a room to "fill in" sound, especially if the room is very deep with respect to its width as indicated in *Figure 4-11c.*
- *Don't* place additional speaker systems in the back of the room so that the speaker systems point toward the front as shown in *Figure 4-11d.* This can severely reduce the intelligibility of the sound. As a rule of thumb, the multiple sets of speaker systems should point in the same general direction. (The exception to this is monitor speaker systems, which are placed on the stage and point back to the performers.

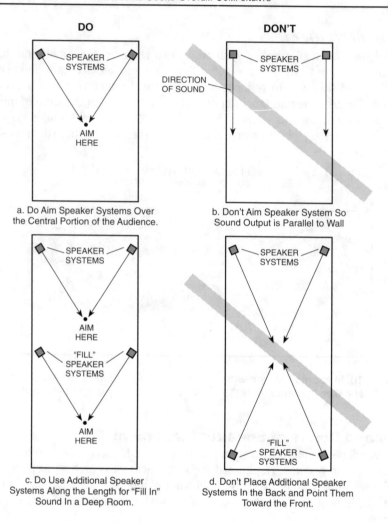

Figure 4-11. The basic do's and don'ts of speaker placement.

When first setting up a sound system, it may be beneficial to mount the speaker systems only temporarily in the locations you think are best suited. Try the sound system (preferably with people in the room) and sit at various places in the audience area. Note any places where the sound is excessively loud or weak. Reposition the speaker systems as necessary. Chapter 6 provides more detailed information for testing the sound system.

MOUNTING THE SOUND SYSTEM COMPONENTS

Sound system electrical components (amplifier, mixing console, etc.) usually are available in either of two cabinet designs: tabletop or rack-mounted. Tabletop cabinets are normally used for smaller, less expensive components. To install them, you just set the component on a tabletop. Rack-mounted components are designed to be secured in a standard 19-inch rack. However, most rack-mounted components can also be placed on a tabletop if you don't want to use a rack.

Rack Mounting

Two mounting racks suitable for sound system components are shown in *Figure 4-12.* The smaller rack, shown in *Figure 4-12a,* is designed to sit on a table and is big enough to hold a mixing console and, perhaps, a power amplifier. The larger rack, shown in *Figure 4-12b,* is designed to sit on a floor and has enough space for a mixing panel (which goes on top), as well as an amplifier, equalizer, and other components.

a. Rack to Sit On Table   b. Rack to Sit On Floor

Figure 4-12. Two 19-inch equipment racks for sound system components: (a) is for tabletop use; (b) is floor standing.

Cooling

Regardless of how you mount the components, it is always important to provide sufficient ventilation to prevent overheating. Adequate air circulation is a must, especially for the power amplifier, which can be permanently damaged if it gets too hot. Ready-made equipment racks like those in *Figure 4-12* are often designed with ventilation slots to assist in cooling. The back of the cabinet is slotted for ventilation and it can be removed if extra cooling is required.

Custom Cabinets

If you plan to install the sound system components in a rack or cabinet of your design, it is crucial that you provide for cooling. Simple convection cooling can be used if at least 30 percent of the surface area of equipment is exposed to the open air. In fully enclosed racks or cabinets, suitable cooling fans (available at Radio Shack) are necessary to force air past the components.

In a custom cabinet or other enclosure, if you plan to install rack-mounted equipment, be sure that the cabinet is designed to accept and secure the standard 19-inch wide professional sound system equipment. If the equipment is physically available, measure the dimensions of the sound system components; otherwise, obtain the dimensions from specification sheets. Use the dimensions to design the custom cabinets.

Lighting and Additional Space

If you plan on operating the sound equipment during a program — as opposed to merely turning on the power amplifier and walking away — be sure to provide adequate lighting so you can see the controls. If the area around the sound system components is dimly lit, a small directional lamp (like those used for reading in bed) is a good choice. The light is bright enough so you can safely and comfortably see what you are doing, but is not so bright that it annoys the audience.

If you plan on using additional sound inputs from a turntable, CD player, and/or cassette deck, place them near the mixing console so that you have immediate access to them during the program. These components usually are designed for tabletop use, so you need to provide table space for them. Avoid an overcrowded work area; if the area is too small, it will not be comfortable to work in. At least leave enough space for a pad of paper.

CONTROLLING REVERBERATION

Some rooms are naturally more acoustically "live" than other rooms. A live room usually has hard, straight boundaries (walls, floor, and ceiling). The sound easily bounces off these boundaries and causes reverberation. The more reverberation, the livelier the room. A certain amount of reverberation is needed to make the sound seem more natural. Two undesired conditions can exist: too much reverberation or too little reverberation. If there is too much reverberation in a room, the program sound can become unintelligible. If there is too little reverberation, it can be difficult to hear the sound adequately. Lack of reverberation is usually the result of too much sound absorbing material in the room: thick carpets on the floor, soft upholstery on the seats, heavy fabric or drapes on all the walls, and acoustic tile on the ceiling.

So there is a happy medium somewhere between too much reverberation and too little. Too little reverberation is difficult to correct and could be expensive; for example, you might have to remove the carpet or drapes. On the other hand, if the room has too much reverberation, perhaps to the point where the echoes are overpowering, here are some simple steps to reduce the unwanted sound reflections.

- The cheapest anti-reverberation fix is adding drapes or fabric over brick or concrete walls. You don't have to cover all the walls, nor do the walls need to be covered ceiling to floor. You can reduce the most objectionable reverberation by adding drapes at only the front and/or back of the room. Drapes only need to be hung from the ceiling to about four feet from the floor. An inexpensive cotton fabric (like that used for building stage scenery) can be applied either directly to the walls or attached to a lightweight wooden framework that is secured to the walls. The fabric can be painted or left natural, as desired.

- The walls nearest the speaker systems can contribute a great deal to the overall reverberation in a room. We discussed absorbing material briefly in *Figure 4-8.* By installing heavy fabric panels behind the speaker systems as shown in *Figure 4-13,* the sound echoes can be reduced to a reasonable level. However, installing the fabric panels can reduce the overall sound level from the speaker systems. Thus, the sound output is lower. Also, placing the speaker systems too close to the wall and fabric panels can increase the low-frequency response of the sound system so the sound is overly "boomy" or "bassy." You may wish to experiment with temporary fabric panels near the speaker systems to see how they affect the overall sound before they are installed permanently.

■ A more costly approach for reducing reverberation is to add acoustic tile to the ceiling. If air conditioning ducts, water pipes, etc. are exposed, you may wish to install a suspended acoustical ceiling. Covering them not only improves the appearance of the room, but also prevents the significant reflections from the ducts.

■ In a room that is not full to capacity, a wooden or concrete floor tends to reflect a great deal of sound. Carpeting is an effective way to reduce sound echoes, but it can be expensive, especially if the room is large. A basic commercial grade carpet with a low nap is usually adequate. Plush carpeting is neither required nor desired.

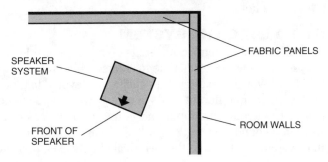

Figure 4-13. Large fabric panels can be placed near the speaker systems in the corner of the room to reduce unwanted reverberation. Care must be exercised that the close proximity of the speaker to the panels does not cause an excess of bass response ("boomy" sound) in the system.

In all cases, be sure to consider the effect of people on the reverberation level of a room. A common mistake is to decide that a room is too "live" when testing a sound system with no audience in the room. The *only* effective test is with an audience, because people absorb the majority of the sound produced by a sound system in an enclosed room. When in doubt, always wait for an actual presentation, complete with audience, to judge the need for applying reverberation control techniques.

MIXING PRO GEAR WITH HOME HI-FI EQUIPMENT

Home hi-fi systems can amplify the same sounds and provide 8-ohm speaker connections like the professional sound system components, and the home system is less expensive. Why not use it, or at least some home components mixed with the pro system?

Mixing home hi-fi equipment with professional sound system components may save you some money, but it's not always a wise choice. Here are some reasons for *not* mixing:

1. Home stereo components usually are not built with the same attention to ruggedness as professional sound system gear.

4

2. Home components may not have the certification required in your area for electrical equipment used in a public building.

3. Signal levels and input impedances can be different between home gear and pro gear. For example, the microphone input of most mixing consoles is either low impedance, or is switchable between low and high impedance. Home hi-fi gear, on the other hand, may be designed to accept only high-impedance microphones. If you use a professional quality microphone, which is almost always low impedance, you will introduce an impedance mismatch and the system will not perform as well.

Still, if you only have one alternative, you can mix turntables, CD players, and cassette decks intended for home use in a professional sound system if you exercise some care. Just be sure to buy the highest quality you can afford, because the equipment likely will get rougher treatment in a public sound system environment than it would in a home.

READY TO WIRE THE SYSTEM

With the main sound system components — mixing console, power amplifier, and speaker systems — in place, it's time to connect them together. Wiring the components is highly involved, so a separate chapter is devoted to this subject. In Chapter 5, you will learn about different kinds of cables, wires, and connector types and the importance of matching the input and output impedances of the various components used in a sound system. You will also learn how to assure safe operation of the sound system by checking the electrical outlets used to power the system.

Wiring Sound Systems ⑤

INTRODUCTION

While the mixing console, power amplifier, and speaker systems form the backbone of the typical sound system, the wires that connect everything together function as the central nervous system. The wiring of a sound system requires careful thought and attention to detail. For example, each wire must be the proper type and use the proper connector. Wire lengths should be kept to a minimum. The signal wires should be physically separated from power cords to reduce the chance of picking up and amplifying 60-Hz power frequency noise commonly called "hum."

This chapter addresses the hows and whys of wiring the sound system. You'll learn which connectors are required for the job. You'll learn the best ways to route wires so they're out of the way of people and things. You'll learn tricks of the trade, such as various ways to wire speaker systems, how to wire a constant voltage distribution public address system, and how to connect everything to assure the cleanest, most static-free sound. You'll learn how to test the AC electrical wiring to ensure proper and safe operation of the sound system. Such tests can be performed in a manner of minutes with inexpensive equipment.

UNDERSTANDING DIFFERENCES IN WIRE AND CABLE

Throughout this book, we use "wire" as a generic term. A wire is basically a length of metal conductor surrounded by an insulating material, usually plastic or rubber. In sound system work, "cables" are more often employed. A cable is composed of one or more individual wires, fused or sheathed together. Most often, the cable is terminated using a unique connector. The type of connector is usually (but not always) determined by the application. A connector for a microphone cable is different than a connector used for hooking up a CD player to a mixing console.

Figure 5-1a shows speaker wire. Even though there are two wires and their insulations are fused together, it is called speaker wire — not speaker cable. The conductors are large (various gauges are available) to ensure minimum power loss between the amplifier and speaker systems, as we discussed in Chapter 4. *Figure 5-1b* shows typical construction of shielded cable used in sound systems. It consists of insulated wires bundled together inside a metal shield covered by an outer jacket of insulation. The shield helps reduce interference from external electrical fields — AC power, static, or other electrical interference — which may be amplified by the rest of the sound system. The metal shield is connected to the chassis ground of the sound system. With the exception of the speaker wires that connect between the power amplifier and the speaker systems, all cabling in a sound system should be the shielded type.

NOTE

Check the appropriate electrical codes for the types of cable allowed. In some locations, a fireproof or fire resistant type is required in some commercial buildings.

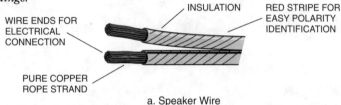

a. Speaker Wire

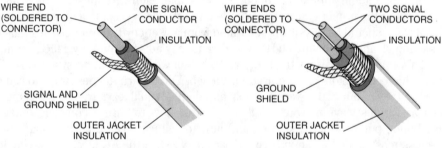

b. Shielded Cable

Figure 5-1. Wire is typically composed of one or more individual conductors, with each enclosed in its own insulation, and is not terminated by a connector. Cables usually contain two or more insulated conductors, with or without an outer shield, and are terminated by a connector.

TYPES OF CONNECTORS AND THEIR USE

There are three major types of connectors used in the typical sound system: 1/4-inch phone, phono (commonly called "RCA connector"), and XLR. Each of these types is discussed in more detail below. For all three types, the male portion of the connector is referred to as the "plug," and the female portion is referred to as the "jack." The plug and jack together form the complete connector.

1/4-Inch Phone Connector

The 1/4-inch phone plug, shown in *Figure 5-2,* and its mating jack are the most common connectors used for sound systems. The plug consists of two pieces of metal (three if the connector is for stereo) separated by a plastic dielectric. The piece of metal at the end of the plug is called the *tip,* and is generally used for the positive side of the circuit. If a 2-conductor, the rest of the plug is called the *sleeve* and is most often used as ground for the circuit. If a 3-conductor, there is an additional insulated *ring* between the tip and the sleeve. The mating jack has two pieces of springy metal to make contact with the tip and sleeve (or tip and ring if a 3-conductor) of the plug.

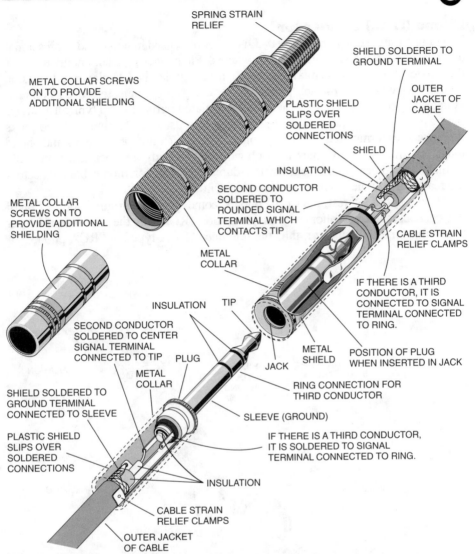

SPRING STRAIN
RELIEF

SHIELD SOLDERED TO
GROUND TERMINAL

METAL COLLAR SCREWS
ON TO PROVIDE
ADDITIONAL SHIELDING

OUTER
JACKET OF
CABLE

PLASTIC SHIELD
SLIPS OVER
SOLDERED
CONNECTIONS

SHIELD

INSULATION

SECOND CONDUCTOR
SOLDERED TO
ROUNDED SIGNAL
TERMINAL WHICH
CONTACTS TIP

METAL COLLAR
SCREWS ON TO
PROVIDE ADDITIONAL
SHIELDING

CABLE STRAIN
RELIEF CLAMPS

METAL
COLLAR

IF THERE IS A THIRD
CONDUCTOR, IT IS
CONNECTED TO SIGNAL
TERMINAL CONNECTED
TO RING.

INSULATION   TIP

SECOND CONDUCTOR
SOLDERED TO CENTER
SIGNAL TERMINAL
CONNECTED TO TIP   PLUG

METAL
SHIELD

POSITION OF PLUG
WHEN INSERTED IN JACK

METAL
COLLAR

JACK

RING CONNECTION FOR
THIRD CONDUCTOR

SHIELD SOLDERED TO
GROUND TERMINAL
CONNECTED TO SLEEVE

SLEEVE (GROUND)

PLASTIC SHIELD
SLIPS OVER
SOLDERED
CONNECTIONS

IF THERE IS A THIRD CONDUCTOR,
IT IS SOLDERED TO SIGNAL
TERMINAL CONNECTED TO RING.

INSULATION

CABLE STRAIN
RELIEF CLAMPS

OUTER JACKET
OF CABLE

Figure 5-2. Shielded 1/4″ Phone Plug and Jack. Two-conductor connections are made to tip and ground through sleeve. Three-conductor connections are made through tip, ring, and ground through sleeve.

It doesn't take a rocket scientist to realize that the 1/4-inch phone connector get its name because it's approximately a quarter of an inch in diameter. The "phone" part comes from its early application as a connector for manual telephone switching systems. Of course, these connectors are no longer used in telephone systems, but the "phone" name remains. In fact, "tip" and "ring" connections are very common in telephone circuits today.

Most microphones come with a 1/4-inch plug (or can be easily modified to use one). As a result, 1/4-inch phone jacks are commonly used for microphone inputs on mixing consoles. But they can be found most anywhere. In fact, they may be the only connectors used, except for the speaker system power amplifier connections.

Phono (RCA) Connector

The phono plug, shown in *Figure 5-3*, is commonly used in home and professional audio systems, which includes sound systems. Phono jacks are commonly used for the line-level inputs to the mixing console; for example, the CD player or turntable inputs. As with the 1/4-inch phone connector, phono connectors used in sound systems are comprised of two electrically insulated parts: the tip typically carries the positive side of the circuit and the sleeve is used for ground. Phono jacks require cables terminated with phono plugs; you cannot mix phone jacks and phono plugs, or vice versa. Cables with phono plugs already attached are commonly available or you can make your own. Adapter between common cable connectors are readily available from most electronics outlets.

As a point of interest, the word "phono" comes from the original application of this connector in early phonographs and radios. RCA was the first manufacturer to use this kind of connector, thus the reference "RCA-type" or "RCA phono."

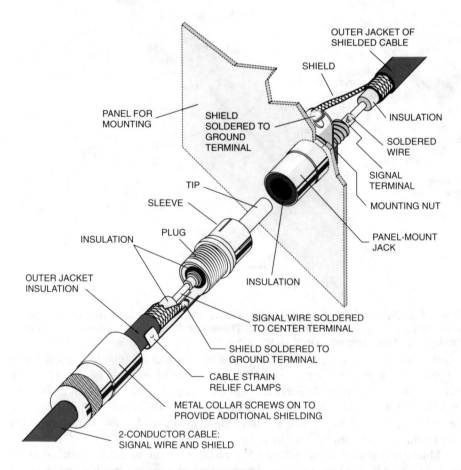

Figure 5-3. Shielded Phono (RCA) Plug and Jack. Two-conductor connections are made to center signal terminal and ground through shield.

XLR Connector

The XLR plug, shown in *Figure 5-4*, is specially designed for professional audio applications. The connector is physically the largest of the bunch, and consists of a heavy metal shell with three conductors inside. It is designed to greatly reduce the chance of interference being picked up by the connector and cable. The XLR connector achieves this by using three conductors—two insulated signal conductors surrounded by a metal shield—as shown in *Figure 5-4* and the balanced connection of *Figure 3-9*. If noise is introduced in the cable, it induces the same signals in the two signal conductors. Because the active signals on the signal conductors are 180° out of phase, the noise signals cancel at the input.

In addition to the two signal conductors, the XLR connector can optionally use a separate shield wire from the ground wire. This further helps to reduce the chance of static and interference from entering the cable and messing up the clear sound from your sound system.

You typically find XLR jacks only for the microphone inputs, which are the ones most sensitive to noise interference. Unlike 1/4-inch phone and phono cables, you are not as likely to find ready-made cables with XLR connectors already attached. You must make these cables yourself. We will describe how shortly.

The XLR connector is the most complex of the three connectors, and, therefore, is the most expensive. It is used mostly on systems for professional industrial use because of additional benefits. First, the plug is designed to secure with the mating jack with a latching mechanism. Under normal circumstances, the plug will not release from the jack unless the release lever is pushed. This helps prevent the plug from accidentally disconnecting from the jack. Second, the ground connection is made first, followed by the signal connections. With the sound system on, this helps reduce the pop that can be heard when connecting and disconnecting the plug. Such pops are not only annoying, but also if the output level of the amplifier is set high enough, it can lead to premature failure of the amplifier and the speaker systems.

MAKING YOUR OWN CABLES

Your sound system interconnecting cables may require a plug on one end and a jack on the other, or the same type connectors on both ends. You can purchase most of the cables with connectors and without connectors, and the separate connectors, at an electronic parts store. Cables come in varying lengths, and it is best to choose the cable that is just long enough for the job. Cables that are too long are cumbersome, and cables that are too short can be a system or safety hazard.

To make the cable yourself by soldering the wire ends to the appropriate connector, you need a soldering pencil with a pointed tip like the one shown in *Figure 5-5*, as well as rosin core solder suitable for electronics work. *Do not use acid core solder or silver solder*. A soldering pencil is recommended because a soldering iron applies too much heat which may damage the cable and connectors.

Getting Ready for Soldering

Attaching connectors to cable ends is not particularly difficult, but it does require some patience. Certain connector designs also require some skill and experience if you expect a good job. It you're a beginner, you may wish to practice before you

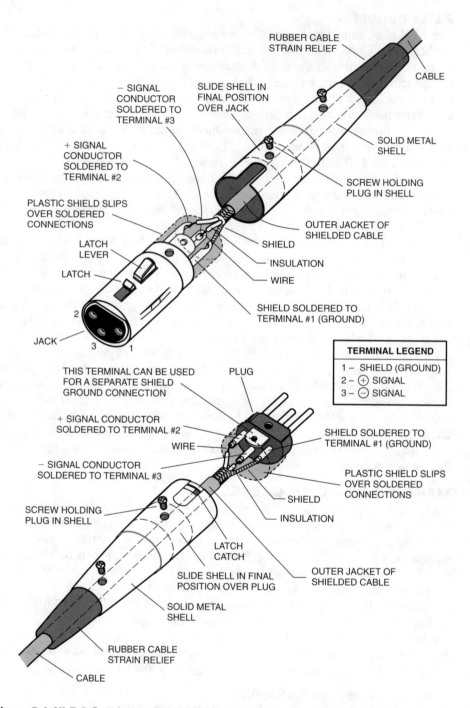

RUBBER CABLE
STRAIN RELIEF

CABLE

− SIGNAL
CONDUCTOR
SOLDERED TO
TERMINAL #3

SLIDE SHELL IN
FINAL POSITION
OVER JACK

SOLID METAL
SHELL

+ SIGNAL
CONDUCTOR
SOLDERED TO
TERMINAL #2

SCREW HOLDING
PLUG IN SHELL

PLASTIC SHIELD SLIPS
OVER SOLDERED
CONNECTIONS

OUTER JACKET OF
SHIELDED CABLE

LATCH
LEVER

SHIELD

INSULATION

LATCH

WIRE

2

SHIELD SOLDERED TO
TERMINAL #1 (GROUND)

JACK

3   1

TERMINAL LEGEND

1 – SHIELD (GROUND)
2 – ⊕ SIGNAL
3 – ⊖ SIGNAL

THIS TERMINAL CAN BE USED
FOR A SEPARATE SHIELD
GROUND CONNECTION

PLUG

+ SIGNAL CONDUCTOR
SOLDERED TO TERMINAL #2

SHIELD SOLDERED TO
TERMINAL #1 (GROUND)

WIRE

− SIGNAL CONDUCTOR
SOLDERED TO TERMINAL #3

PLASTIC SHIELD SLIPS
OVER SOLDERED
CONNECTIONS

SCREW HOLDING
PLUG IN SHELL

SHIELD

INSULATION

LATCH
CATCH

SLIDE SHELL IN FINAL
POSITION OVER PLUG

OUTER JACKET OF
SHIELDED CABLE

SOLID METAL
SHELL

RUBBER CABLE
STRAIN RELIEF

CABLE

Figure 5-4. XLR 3-Conductor Shielded Plug and Jack. Three-conductor connections are made to plug and jack with signal conductors connected to terminals 2 and 3 and the shield connected to terminal 1.

Figure 5-5. Use a soldering pencil and stand to make your own cables and connector ends. The soldering pencil should be rated at 30 watts or less, and the tip should be designed for general-purpose or printed circuit board use.

attempt to solder a cable connection. In any case, go slowly: a hurried job almost always results in a poor job.

The steps for soldering cable wires to connectors are shown in *Figure 5-6*. *Figure 5-6a* shows a shielded 3-conductor cable (two signal conductors and shield). The cable has an outside sheath or jacket of insulation. Use a sharp utility knife to cut the outer jacket and the shield back as shown to expose the insulated wires. Do not cut off so much of the outside sheath that the individual wires of the cable are exposed, as shown in *Figure 5-6e*, when the connector is assembled.

Before you can solder the cable wires to the connector, you must trim off the insulation covering the wire end. Use the dimensions in *Figure 5-6a* as a guide. For best results, use a wire stripper tool that has cutters for a particular wire gauge or lets you "dial in" the gauge of the wire. This helps speed up the work, and keeps you from accidentally stripping away part of the wire with the insulation.

The signal wires and the shield should first be lightly "tinned," as shown in *Figure 5-6b*, before actually soldering the connection to the connector terminals. Tinning involves applying a *very light* coat of solder to the wire. To tin the wire, touch the soldering pencil to the wire. Wait a few seconds for the wire to get hot, then lightly touch the solder to the wire (don't touch the solder to the soldering pencil). Properly tinned conductors should look bright and shiny.

Most connectors use either of two methods for soldering wires: the "solder pot" and the eyelet. A solder pot is simply a sleeve of metal with a well or "pot" in the end. You insert the wire into the sleeve, then apply solder to fill the pot. With the eyelet type, you wrap the end of the wire around the eyelet, producing a secure connection. Examples of both solder pot and eyelet are shown in *Figure 5-6e*. Prior to soldering the wires, be sure to insert the connector collar and inner insulating sleeve over the cable.

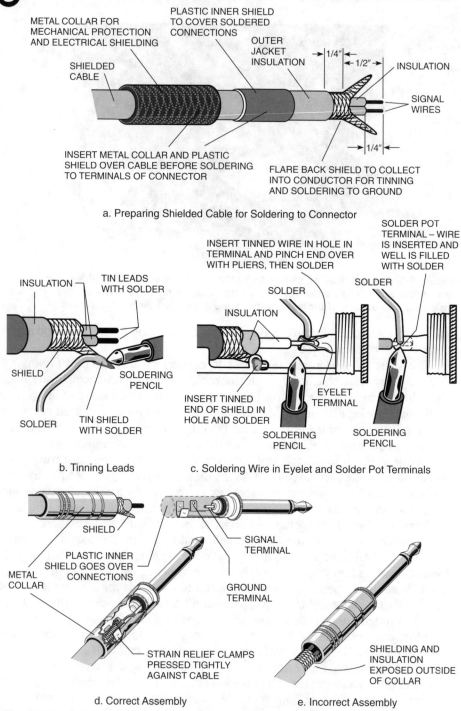

a. Preparing Shielded Cable for Soldering to Connector

b. Tinning Leads

c. Soldering Wire in Eyelet and Solder Pot Terminals

d. Correct Assembly

e. Incorrect Assembly

Figure 5-6. Soldering Cable Wires to Connectors. Shielded cables must be prepared for soldering and assembly by trimming insulation and the shielding, tinning the leads, and inserting the wires in eyelet or solder pot terminals. Prior to soldering wires, slip metal collar and plastic inner shield over cable.

Because of the small size of the connector and cable, it is best to secure both parts in a vice or holder specially designed for electronics works, (RS 64-2093 called "Helping Hands" is an example) where a "third hand" securely holds the two parts to be soldered. The final assemblies of phone, phono, and XLR connectors are shown in *Figures 5-2, 5-3,* and *5-4,* respectively.

Basic Soldering Techniques

Before soldering, apply solder to the hot tip of the soldering pencil to tin the tip, then wipe off the excess solder with a damp cloth or sponge. While soldering, wipe the tip of the soldering pencil as necessary to keep it clean. When the tip loses its shiny appearance, re-tin it. When the tip becomes damaged or worn, replace it.

The soldering pencil is used to heat the work — the wire end and the connector — while applying solder to the work. Do not apply solder directly to the soldering pencil and let it flow onto the work. You might end up with a "cold" solder joint which doesn't adhere well to the metal surfaces of the connector, so electrical connection is impaired. Do not apply heat any longer than necessary. The tip of a soldering pencil is between 650 and 750 degrees Fahrenheit which, if applied too long, can melt the insulation off the wire. A good rule of thumb is that the soldering pencil should not be on any one spot for more than about five seconds.

Apply only as much solder to the joint as is required to coat the connection. Applying too much solder can lead to a solder glob that could cause a short circuit. Once the solder flows uniformly around the joint (and some will flow to the soldering tip), remove the solder and soldering pencil and let the joint cool. Do not move the wires and connector as the solder cools and solidifies; if you do, a cold solder joint and poor electrical connection might result.

Finishing the Cable

Many connector plugs are designed with a crimp type stain relief that helps prevent the soldered connection from coming apart if the cable is yanked out of the jack by the wire rather than gently pulling on the plug. To use the strain relief, use pliers to gently crimp the nib of metal over the cable as shown in *Figures 5-2, 5-3,* and *5-4,* making sure that you don't cut into the shield and wire insulation.

Complete the job by replacing the collar over the connector. If the collar is made of metal, be sure the inner insulating plastic, rubber, or paper shield is in place to prevent the wires from contacting the collar and causing a short circuit.

ROUTING CABLES BETWEEN COMPONENTS

With cables of the correct length, you are now ready to route them between the components of your sound system. The routing for a permanent installation is usually different from that of a temporary installation.

Permanent Installation of Cables

Permanent installation of sound system cables entails routing the cables so that they are out of reach, and preferably, out of sight. The construction of the room where the sound system is used greatly determines the method of installing cables. Before committing to any permanent installation, you should consult local business codes for special requirements for your area. The three methods described here are depicted in *Figure 5-7*.

- If the auditorium has an accessible basement or crawl way: route the cables through the floor to the basement or crawl way. If local building codes require, place the sound system cables inside plastic or metal conduit. Secure the conduit to joists or the underside of the floor with suitable brackets.
- If the auditorium has a false (suspended) ceiling: route the cables up the wall and through the ceiling tiles. Consult local building codes to determine if the sound system cables must be enclosed in metal or plastic conduit.
- If the auditorium has no basement, crawl way, or suspended ceiling: the cables must be routed along the walls, baseboards, and floors of the auditorium and situated so that people will not trip over them. Use plastic raceways, which are available at home improvement stores, to conceal and protect the cables. By piecing together the raceway using segments, connectors, and corner pieces, you can not only hide the cable from the audience, but also secure it where people are less likely to become entangled with it.

Temporary Installation of Cables

If the sound system installation is temporary — for example, you are setting up a sound system in an elementary school auditorium for the Christmas pageant — there is no need to install the cable in the ceiling, basement, or cable raceways. However, it is vitally important that any cables are kept out of the way so that people will not trip over them. This avoids damaging the cables, as well as a potential lawsuit!

Always keep in mind that people may stumble or trip over exposed cables. Use adhesive tape (duct, masking, etc.), rubber floor strips available at industrial supply houses, and/or wide runner carpet strips to hold and cover cable runs around the room or in heavy-traveled areas. Recall that duct tape may leave a sticky residue which may harm some surfaces.

RUNNING SIGNAL CABLES AND POWER CORDS SEPARATELY

Even though your sound system is connected with shielded cables, electrical noise (hum) caused by the 60-Hz alternating current flowing through a nearby power cord may be heard through the speaker system. You can prevent most induced hum by keeping power cords physically separated from the signal cables. Whenever possible, avoid placing power cords and signal cables side by side. If they must run parallel, keep them apart by at least six inches. Never let loops of power cords and sound cables near one another. The loops act as transformers and a very noticeable amount of hum can result.

UNDERSTANDING BALANCED AND UNBALANCED LINES

Signal lines in a sound system are classified in two broad terms — balanced and unbalanced. Despite the terminology, an unbalanced line is not necessarily worse than a balanced line. Most small sound systems use only unbalanced lines; there is nothing wrong with this, and your system will sound just fine.

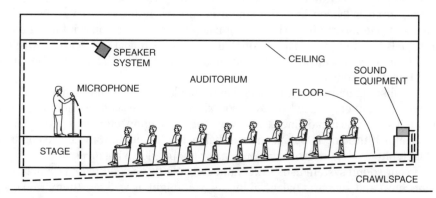

a. Under Floor

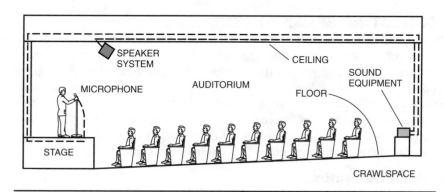

b. Ceiling

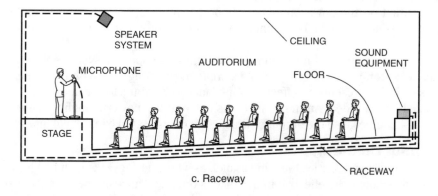

c. Raceway

Figure 5-7. Three methods of permanently routing cables in an auditorium. In all cases, the aim is not only to make the cables as inconspicuous as possible, but also to keep them away from traffic patterns where people may trip over them.

In an unbalanced line, the sound is carried by two wires — a signal wire and ground. In an balanced line, the sound is carried by three wires — two for the signal and one for ground. The two signal wires carry the same information, but the signals are a mirror image of one another. In technical terms, the two signals are 180 degrees out of phase. As you read earlier in this chapter, this technique helps reject spurious noise.

The distinction between balanced and unbalanced lines largely has to do with the type of connector used in your sound system. The XLR connector provides balanced or unbalanced lines. The way the XLR connector is wired internally determines if the line is balanced or unbalanced. In addition, the three-conductor 1/4-inch phone plug is also occasionally used to provide a balanced line.

If your sound system uses XLR connectors (or three-conductor 1/4-inch phone connectors) and supports balanced inputs, you might as well use them. Make sure the connectors are wired as shown in *Figure 5-4* so that they support a balanced signal. The schematic of a balanced signal line was shown in *Figure 3-9*.

CONNECTING SPEAKER SYSTEMS

Recall from Chapter 3 that the power amplifier output must be matched to the effective impedance of the speaker system. Because it is often necessary to use more than one speaker system (per channel) in larger auditoriums to disperse the sound to the entire audience, two or more speaker systems must be wired together. However, since each speaker system has its own impedance, wiring them together changes the effective impedance presented to the power amplifier. The effective impedance can be controlled by the method of electrically connecting the speaker systems. The three connection schemes are shown in *Figure 5-8*.

Series Connection

For speaker systems connected in series, as shown in *Figure 5-8,* simply sum the impedances of all the speaker systems. For example, if you connect two 8-ohm speaker systems in series, the effective impedance is 16 ohms. You then use the 16-ohm output on the power amplifier for proper impedance matching. Four 4-ohm speaker systems could be connected in series for the same 16-ohm effective impedance.

There is a practical maximum to the number of speaker systems that can be effectively connected in series. For example, connecting four 8-ohm speaker systems in series presents an effective impedance of 32 ohms to the amplifier. This is decidedly a bad idea. An impedance mismatch at the power amplifier causes distorted sound because the amplifier is asked to operate at maximum voltage swing.

Parallel Connection

If you connect speaker systems in parallel, as shown in *Figure 5-8,* and all have the same impedance, the effective impedance is the impedance of one speaker system divided by the number of speaker systems. For example, if two speaker systems rated at 8 ohms each are connected in parallel, the effective impedance is 4 ohms. Similarly, if four speaker systems rated at 16 ohms each are connected in parallel, the effective impedance is 4 ohms. The 4-ohm power amplifier output is used for either situation.

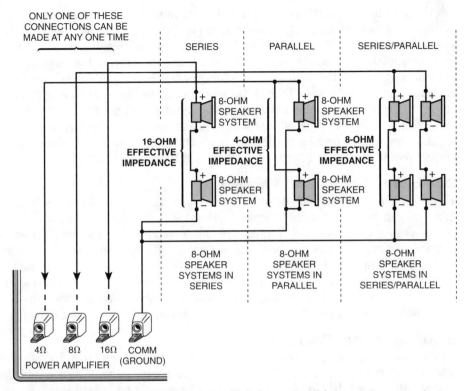

Figure 5-8. Speaker System Connections. The three possible ways to connect speaker systems to the output of a power amplifier. Common ground and only one PA terminal can be used at a time. Be sure to calculate the effective impedance of the speakers no matter what connection scheme you use. For best results, all speaker systems should have the same impedance.

For speaker systems connected in parallel that don't have the same impedance, use the following equation to find the effective impedance (Z_T):

$$Z_T = \frac{1}{\dfrac{1}{Z_1} + \dfrac{1}{Z_2} + \ ... \ + \dfrac{1}{Z_n}}$$

As a general rule of thumb, connecting speaker systems in parallel is the more desirable approach for a number of reasons. First, when connected in series, if any speaker system fails, none of the speaker systems will work because the circuit is completed through *all* of the speaker systems, as indicated in *Figure 5-8*. If the circuit is broken because of a damaged speaker system, no sound will come through any of the speaker systems.

Second, when connected in series, speaker systems tend to interact with one another, which can result in distortion. All speaker systems exhibit a damping action, which is the result of the natural resistance of the voice coil in the speaker system against the permanent magnet located nearby. This damping action, though mechanical in nature, can result in lower signal levels through the speaker system line.

On the other hand, there is a practical maximum to the number of speaker systems that can be effectively connected in parallel. At some point, the amplifier will no longer be able to supply enough power to each of the speaker systems to produce adequate sound. As an example, the maximum current the power amplifier can supply limits the number of speaker systems in parallel. As a general rule of thumb, you are safe if you limit the number of speaker systems to four (8-ohm) or eight (16-ohm) per channel, assuming the speaker systems are full-range, and the power amplifier is rated at 100 watts per channel. For example, four 8-ohm speaker systems in parallel results in an effective impedance of 2 ohms, and a current requirement of greater than 7 amperes at 100 watts output.

Combining Series and Parallel

To get around the limitations of an all series connection, or an all parallel connection, you can connect multiple speaker systems in a combination of parallel and series, as shown in *Figure 5-8*. You can use this technique, for example, to connect four speaker systems together, and still present an effective impedance of 8 ohms to the power amplifier. Two of the speaker systems are wired in series, and the two series combinations are wired in parallel. Assuming all four speaker systems are rated at 8 ohms, the effective impedance for the *array* of speaker systems is 8 ohms, so you would connect the array to the 8-ohm output of the power amplifier.

Insuring Proper Polarity

When connecting multiple speaker systems, it is absolutely vital that you observe proper polarity at all times. *Figure 5-8* shows the proper polarity. If you mix the polarity, the sound delivered by your sound system will be substandard. It may sound excessively "boomy," and the output level may be diminished.

Connecting Speaker Systems in a Constant Voltage Distribution System

The section above explained how to connect speaker systems in a typical sound reinforcement system. Public address (PA) systems typically use a different mechanism for connecting speaker systems. Each speaker system is connected to a special transformer, called a *line transformer*, which is connected to the 70-volt distribution line from the power amplifier. Many power amplifiers designed for PA systems provide both standard 4/8/16-ohm speaker connections and the 70-volt distribution connection.

Figure 5-9 shows a typical PA system using a 70-volt constant voltage distribution system. Note that a separate line transformer is used for each speaker system. Each transformer has multiple taps so that you can control the amount of power reaching the speaker system. Typical taps are 0.62, 1.25, 2.5, 5 and 10 watts. To install a constant voltage PA system, you must solder the wires from the primary of the line transformer to the distribution line and the secondary of the line transformer to the speaker system. Plug-and-jack connectors usually are not used, because PA systems are typically permanent.

One benefit of the constant voltage distribution system is that you don't need to worry about the aggregate impedance of speaker systems you attach to the power amplifier. You can add or remove speaker systems anytime and it will not

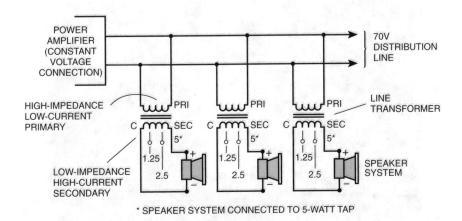

* SPEAKER SYSTEM CONNECTED TO 5-WATT TAP

Figure 5-9. Constant voltage distribution, as used in public address systems, uses tapped transformers to provide the signal to each speaker system on the line. The transformer taps allow you to select a different wattage for each speaker to control the volume level for different areas of the building.

make any difference in the performance of the amplifier or the speaker systems, as long as the maximum number allowed is not exceeded. The total depends on the wattage taps used and the design of the power amplifier. Consult the operator's manual that came with the power amplifier to determine the practical maximum number of speaker systems you can attach to the constant voltage distribution line.

Another benefit is that wire length is not a major consideration as it is with standard sound system speaker system connections. The 70-volt distribution line can be hundreds of feet long (because it feeds high-impedance line transformer primaries), and it will have little effect on the power delivered to each speaker system. Each line transformer is usually mounted very near its speaker system so the wires connecting the low-impedance, high-current secondary to the speaker system are short.

TESTING THE AC POWER FOR YOUR SOUND SYSTEM

There is good reason to test the 117 VAC that powers your sound system. You can spot problems such as polarity reversal that might cause excessive hum, or more importantly, you can determine if the electrical system is wired properly, and therefore, is safe. While it is rare, if the electrical system is wired incorrectly, it is possible to get a nasty shock simply by touching a metal cabinet of a sound system component, or even the metal case of a microphone! You can perform the tests on the electrical circuits using either an AC outlet testing plug or an ordinary volt-ohm meter.

WARNING
*Fatal voltage and current are present at a live outlet. Be sure to follow all safety precautions when testing the outlet. **Do not touch the exposed metal of the testing leads while they are inserted in an electrical outlet.***

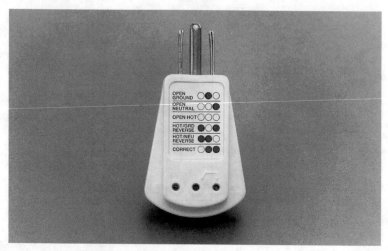

Figure 5-10. An ac testing plug provides a quick and easy indication of any faults in the electrical wiring or outlet.

Using an AC Outlet Testing Plug

The ac outlet testing plug shown in *Figure 5-10* (RS 22-101) is extremely easy to use. Just plug it into the outlet to test. Light-emitting diodes on the plug indicate any fault. If there is a fault, you should have the outlet serviced by a qualified electrician. The two most dangerous faults are reverse polarity and "lifted ground." More about these as you learn how to test for them yourself.

Using a Volt-Ohm Meter

While an ac outlet testing plug is very convenient, it is not necessary to determine if the electrical wiring is faulty. A volt-ohm meter suitable for use with 120 VAC is all you need. The tests that follow assume a three-wire (grounded) polarized outlet, as shown in *Figure 5-11*. If the outlet is wired correctly, the short slot on the right is the "hot" side of the line, and the longer slot on the left is the neutral side. The rounded hole below is earth ground.

Prior to conducting the tests, set the meter to measure VAC. If your meter is not auto-ranging, select a range higher than 120 volts.

Test 1. Insert the test probes into the hot and neutral slots. (It doesn't matter which probe goes in which slot.) You should get a reading of about 117 VAC (it may range from 111 to 123 VAC). If you don't, the outlet may not be powered, or the test probes for the meter may not be making contact inside the outlet.

WARNING

If the meter reads 220 VAC when conducting Test 1, it could indicate that the neutral wire is "lifted," and the outlet is receiving both sides of the incoming 220 volts from the circuit breaker panel. **This must be corrected immediately, as a serious hazard exists.**

Test 2. Insert the test probes into the hot and ground slots. You should get a reading of about 117 VAC.

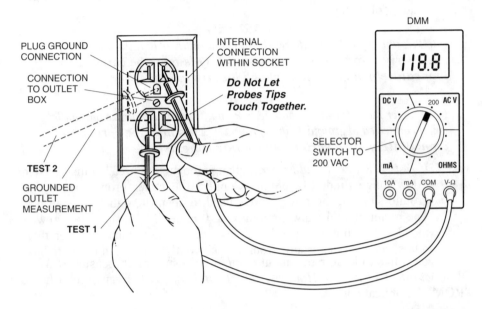

a. Test 1 and 2

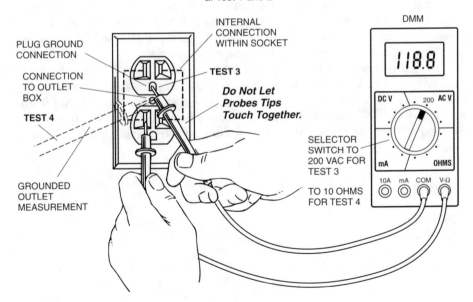

b. Test 3 and 4

Figure 5-11. You can use a volt-ohm meter to perform these four tests on each outlet to be used by the sound system. Any deviation from the described test results may indicate a serious shock hazard.

Test 3. Insert the test probes into the neutral and ground slots. You should get a reading of 0 volts.

WARNING

If tests 2 and 3 yield opposite results, it could indicate that the hot and neutral wires are reversed inside the outlet. **This must be corrected immediately, as a serious shock hazard exists.**

WARNING

If the meter reads 0 volts for both tests 2 and 3, it could indicate that outlet is not connected to ground ("lifted ground"). **This must be corrected immediately, as a serious shock hazard exists.**

Test 4. *If Tests 1, 2, and 3 read properly,* change the meter to read ohms and select a low range (under 1k ohms). Connect the meter probes between the ground slot and the center screw used to keep the outlet cover in place. The meter should show continuity — 0 ohms. Make sure the screw is a metal screw, not a plastic insulating one. If the reading is other than 0 ohms, check to make sure ground wire in outlet box is connected to box.

If you'd like to learn more about using a volt-ohm meter, consult *VOM and DVM Multitesters for the Hobbyist and Technician,* A.J. Evans, published by PROMPT Publications.

SUMMARY

This chapter discussed using cables and connectors to wire your sound system, the differences between wires and cables, the three main types of connectors used in sound systems (1/4-inch phone, phono, and XLR), and stressed the importance of using shielded cables for carrying sound system signals.

You learned how to make your own cables by soldering cable wires to connectors, how to route cables for permanent and temporary installations, and the various ways to connect speaker systems to a power amplifier. Finally, you learned how to test the AC outlets used by your sound system to ensure safe operation. In the next chapter, we provide details for testing, adjusting and operating the sound system.

Testing, Adjusting, and Operating the Sound System

INTRODUCTION

Sound systems are a lot like live television of the 50s and 60s: the audience is instantly aware of any mistakes, and there's no way to go back for a "retake"! Sound system events are live and spontaneous; because of this, you achieve the best and most professional sound through careful setup and planning. This requires testing and adjusting the sound system prior to admitting the audience.

When the actual performance starts — whether it's a music program, a stage play, or a lecture — the operator of the sound system is in complete control of all the sound system variables. In the well-planned sound system, problems and mistakes are usually rare.

This chapter details testing and adjusting the sound system when an audience is not present, then operating the system with an audience present. You'll learn how to place the microphones for best overall pickup, how to adjust sound levels, and how to anticipate the effect of an audience on the sound level and quality.

When operating the sound system in a "live" setting, you'll learn how to carefully adjust levels, equalization, and balance so that the audience is not aware of the changes. The chapter also discusses adding special effects, and mixing the sound for recording or broadcast.

TECHNIQUES FOR PROPER MICROPHONE PLACEMENT

Of all the tasks of setting up and operating a sound system, placing the microphones is one of the most important, if not *the* most important. Placing the microphone in the wrong location — as well as using the wrong kind of microphone — can ruin an otherwise excellent sound system. No mixing console, power amplifier, or speaker system can compensate for the bad sound of a misplaced or incorrect microphone.

The type of production greatly influences the type and location of microphones. We'll consider the best placement schemes for lectures and speeches, musical acts, and theatrical performances. Most sound system use falls within one of these categories.

Lectures and Speeches

For lectures and speeches, a single unidirectional microphone placed directly in front of the speaker, as shown in *Figure 6-1,* is the best choice. If the speaker uses a lectern, the microphone can be located on the lectern, but should be acoustically

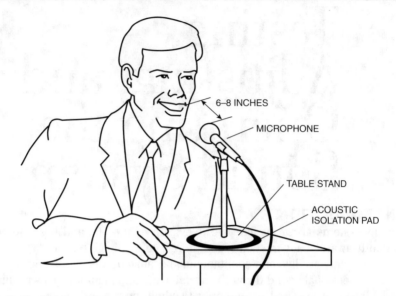

Figure 6-1. Optimal microphone placement for a speaker at a lectern is to place the microphone between 6-8 inches from the speaker's mouth. The microphone should not obstruct the audience's view of the speaker.

isolated from it so that it does not pick up the sound of the speaker's hands touching the edges. Otherwise, you risk picking up rasps, scrapes, knocks, and other unwanted effects.

Don't use a large microphone that blocks the speaker's mouth. Consider using a smaller microphone, or a "lapel" microphone, that is worn by the speaker.

Musical Acts

The topic of proper microphone placement for musical groups is a complex one and can be difficult, especially if there are many singers or acoustic instruments; however, here are some general techniques:

- For one singer and an acoustic guitar (a common arrangement), place one microphone 6-12 inches from the singer's mouth, and another microphone 9-18 inches from the sound hole (or "mouth") of the guitar as indicated in *Figure 6-2*. Extra space is needed between guitar and microphone so the singer doesn't accidentally bump the microphone with the guitar, which, if it happens, can be quite annoying to the audience!
- For a instrument soloist, such as a violinist, trumpeter, or saxophonist; a single cardioid microphone is often sufficient. The microphone is used to pick up both the instrumentalist's voice, if he or she speaks to the audience, and the instrument. The instrumentalist should be allowed to dictate the location and distance from the instrument and the microphone.
- For two singers, each with an acoustic guitar, a useful arrangement is an omni-directional microphone (or stereo cardioid microphone) placed between the two singers. Additional microphones are placed near the two guitars.
- For a single piano, place one microphone 12-18 inches from the pianist (whether or not the pianist sings), and one microphone near the sound board of the piano.

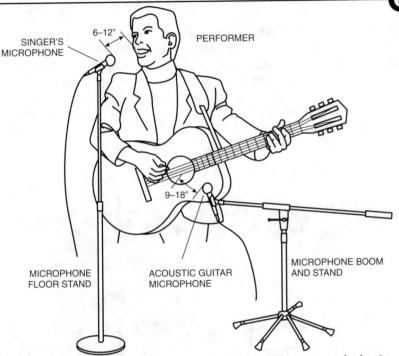

SINGER'S MICROPHONE

6–12"

PERFORMER

9–18"

MICROPHONE FLOOR STAND

ACOUSTIC GUITAR MICROPHONE

MICROPHONE BOOM AND STAND

Figure 6-2. Optimal microphone placement for a single-performer musical act, using an acoustic guitar.

For console pianos, place the microphone behind the piano; for grand pianos, open the top cover of the piano and point the microphone inside (but don't place it all the way inside).

■ For a group with multiple singers and players, place a single microphone 6-12 inches from the singers. If you have additional microphones, place them near the instruments. If this is not possible because of lack of microphones or microphone inputs on the mixing console, place the members of the group as close together as possible so they can share microphones.

Of course, for a group with electrified instruments, there is no need to place microphones near the instruments. Instead, connect the pickup of the instrument — such as a magnetic pickup from an electric guitar — to the proper input of the mixing console. Multiple mixers may have to be coupled together to provide all the necessary microphone inputs.

Note that singing and playing styles differ greatly between people. Some singers are not strong vocalists, and, therefore, need a microphone placed as close to their mouth as practical for adequate pickup. In such cases, you will need to devote a microphone to the one singer. At one time, it was an accepted musical style for two or more singers to sing very closely together and share one microphone; however, today a separate microphone is usually provided for each singer. If possible, make sure the musical group rehearses at least one song using your sound system, so that you can better anticipate their special needs. Be on the lookout for overly strong or weak singers, as well as members who play their instruments very softly or very loudly. Use your best judgment in placing the microphones for best overall sound pickup.

Theatrical Performances

Most theatrical performances cannot be conducted with microphones on the stage. Rather, when microphones are used, they are typically placed above the stage, or on rods near the front edge of the stage, to pick up all the actors, as shown in *Figure 6-3.* In outdoor theaters, or productions where it is impossible to place a microphone over the stage, mikes can often be placed creatively on the set, out of view of the audience. Use caution so that these microphones don't pick up the sounds of the actors walking or using the furniture.

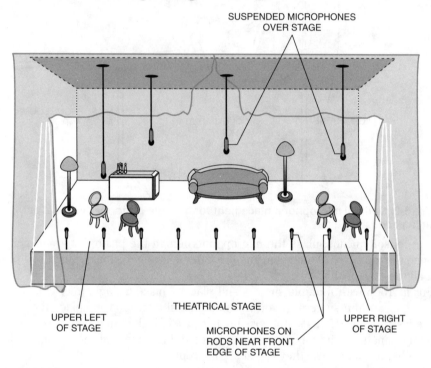

SUSPENDED MICROPHONES
OVER STAGE

THEATRICAL STAGE

UPPER LEFT
OF STAGE

MICROPHONES ON
RODS NEAR FRONT
EDGE OF STAGE

UPPER RIGHT
OF STAGE

Figure 6-3. Optimal microphone placement for a theatrical performance is to suspend the microphones over the stage, or place them on small rods near the front edge of the stage. The microphones should not be in view of the audience. If this cannot be avoided, the microphones should be painted black to make them as unobtrusive as possible.

Another choice is to use a PZM™ microphone located on the stage itself. A PZM microphone uses a different sound pick up technique than a conventional microphone. It is expressly designed to pick up sounds in all directions equally well up to distances of 20 to 30 feet. Though designed for pickup of individuals seated around a large conference table, the PZM microphone is also well suited for many small-scale theatrical performances. For best results, consider using two PZM microphones, as shown in *Figure 6-4,* located near the upper-right and upper-left areas of the stage. Possibly locate the microphone and its cord behind a plant or lamp so it is out of the way of the actors, but is not restricted in its sound pickup.

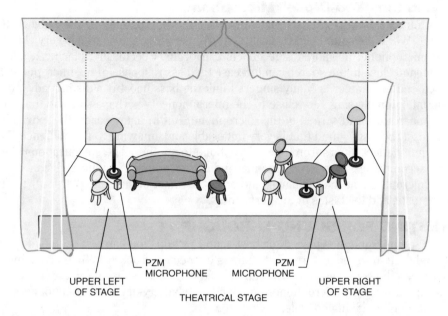

Figure 6-4. The PZM microphone is designed to pick up sounds in a large area and is ideal for theatrical performances.

Wireless Microphones

Depending on your budget, another alternative is to attach a wireless microphone to each performer. This is often done in musical performances, especially in large theaters with a live orchestra in front of the stage. Each performer must be equipped with a separate wireless microphone, and each microphone transmitter must be tuned to a different frequency. Tuned receivers located near the mixing console pick up the signals from each microphone. Professional-quality miniature wireless microphones can cost in excess of a thousand dollars per set, so obviously, this is not an option many sound system operators and local theater producers can afford.

Placing the Microphones On Stands

Except for lapel microphones that attach directly to a person, you will need to place each microphone on a stand and adjust the stand for final placement and distance from the source.

For singers and lecturers, a simple floor stand is usually adequate. The microphone attaches to the stand with a slip-on adapter. This allows the microphone to be easily removed and replaced on the stand.

For instruments, some type of "boom stand" is generally required. The boom allows you to place the microphone close to the instrument, but still not interfere with the performer. The boom is available separately and attaches to most standard microphone stands. On one end of the boom is the slip-on adapter for attaching to the microphone; on the other end is a counterweight to help balance the boom. See *Figure 6-2* for an example.

Avoid the "Too-Close" Microphone

A common misconception is that the closer the microphone to the subject, the better. While the sound pickup is always stronger when the source is very close to the microphone, unwanted side effects can occur. Specifically, placing a microphone too close to the source can increase the bass response of the microphone as discussed in Chapter 3. Many singers think the bass boost is a distinct advantage and intentionally sing very close to the microphone — perhaps so close that their lips touch the wind screen of the microphone. But in many cases, the extra bass causes a "boomy" sound that is very noticeable and annoying to the audience. The proximity effect differs from one microphone to another. The specifications sheet that accompanies the microphone may indicate a "nominal" or "best" distance. Try the microphone at that distance, and adjust it (most often from 4" to 12") to suit your taste, and the tastes of the performers.

TESTING FOR SOUND THROUGHPUT

With the microphones placed for best sound pickup, you can test them with your sound system to make sure everything is connected properly. Whether testing or operating the sound system, set it up by following these steps in the order given:

1. Turn on any microphones equipped with battery-operated preamplifiers.
2. On the mixing console:
   a. Adjust the input levels for all microphones and other inputs to minimum.
   b. Adjust the output level to minimum.
   c. Turn off all special effects, such as delay and echo.
   d. Turn on the mixing console.
3. On the power amplifier:
   a. Adjust the level control (if so equipped) to minimum.
   b. Turn on the power amplifier.
   c. Adjust the level control (if so equipped) to about 50-60 percent — or maybe there is a level you have previously identified.
4. On the mixing console:
   a. Adjust the output level to about 60-70 percent.
   b. Test each input by switching it on (if equipped with switches) and adjusting the level to about 60-70 percent. Speaking *softly* into a microphone provides a quick sound test.

Notice in the steps above that you adjust the level to about two-thirds rather than maximum. This allows for a bit of headroom and adjustment range. You can always lower the level of an input or output, but you can't increase it if it's set up at maximum. If you find yourself always pushing your sound system to the limits — the amplifier all the way up, as well as all the inputs and outputs — it could be an indication that the system is underpowered for its application. Consider upgrading the system, or at least critical parts of it, such as the microphones or power amplifier.

The Sound Check

Once you have tested all the microphones and other inputs, you can set the initial gain of the system by performing a "sound check." In this check, you adjust the levels of all the microphones and other inputs (such as CD player, tape deck, or turntable). If at all possible, perform this sound check with the people that will be on stage — the lecturer or performer(s). If these people are not available, have another person speak into the microphones, or provide input from instruments, so that actual sound inputs are used. This allows you to adjust the levels of the sound system to match the actual program material. The 60% amplifier output level and the 60% master level of the mixing console may be unnaturally high because the room is empty. Temporarily decrease the output level of the mixer first, and then the amplifier, then do the sound check again.

If you are using more than one microphone or other input, carefully adjust the input levels for a natural sounding balance. For musical programs, the vocals should be clearly heard over the instruments, but the instruments should not be so quiet that the audience has difficulty hearing them with the singing. Program mixing is an art in itself, and only experience can guide you.

If the program is varied — different musical acts, for example — you may need to adjust the mixing console for each portion of the program. For example, in the first portion of the program, you may need to increase the level of microphone A because the singer or speaker does not have a strong voice. But for the second portion of the program, you may need to decrease the level. In such instances, it is often helpful to cut small strips of masking or drafting tape, and place them beside the input sliders on the mixing console as indicated in *Figure 6-5*. Use pen or pencil to mark the preset positions for each portion of the program. You might even consider using color-coding since the lights may be turned down during the program so that it is difficult to read letters on the tape.

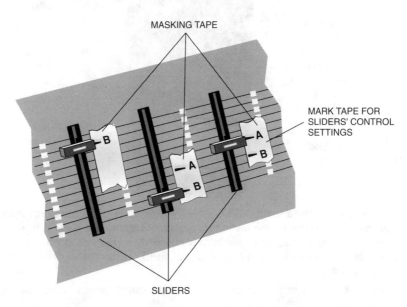

Figure 6-5. Use strips of masking tape to mark predetermined settings of mixing console slider controls. Use color-coding if lettered markings will not be visible during a performance.

Checking Listening Levels

During the sound check, after the mixing console has been set, it's a good idea to check the sound levels throughout the room. You can do this by slowly walking to different points of the room, and listening for any obvious "nulls" or "peaks" in the sound level. If the difference in sound level is obvious to your ear, you may wish to rearrange the speaker systems or the seating, or add appropriate sound baffling material to control the sound. We discussed the effects of sound in an enclosed room in Chapter 2.

While the "ear test" is satisfactory for most small sound system installation, a surer method for testing the sound levels in a room is to use a sound-level meter, such as the one shown in *Figure 6-6*. Sound level meters are inexpensive testing devices designed for only one purpose; that is, to indicate the relative volume of sound at a given spot. The sound level meter consists of a sensitive microphone connected to a precision amplifier. The amplifier output is connected to an indicating device; either an analog meter or a digital readout, which indicates the strength of the sound picked up by the microphone in dB SPL (recall that SPL stands for sound pressure level).

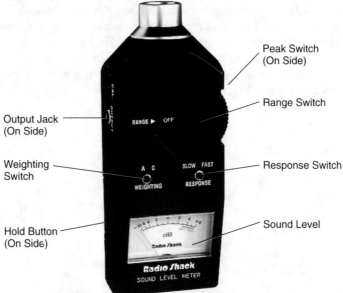

Figure 6-6. The sound level meter allows you to take accurate measurements of sound levels. Use the range switch to select the appropriate dB SPL range. Use the weighting and response switches to set the desired parameters for the measurement. Read the sound level (in dB SPL) on the meter display. (Courtesy of Radio Shack.)

Sound level meters also incorporate various controls so that you can tailor the measurement.

- Range Switch — Adjust the "range switch" to the highest peak you anticipate; for example, 70 dB if you expect no sound over 70 dB. Setting the range to just about the maximum output level improves the accuracy of the meter.

- Weighting Switch — A weighting switch lets you switch between "A" and "C" weighting. "A" weighting closely approximates the hearing characteristics of the human ear, where higher frequencies are heard more readily than lower frequencies. "C" weighting is a more linear measurement, where all the frequencies are considered equally. In most cases, switching to "C" weighting will result in a slightly higher reading because the lower frequencies are not ignored. The type of program influences the weighting that should be used. Rock music is often best measured using the "C" weighting; other programs can be measured using "A" weighting.

- Response Time Switch — A "response time" switch lets you adjust the rate at which the meter takes its measurements. A fast response time indicates the instantaneous sound level. A slow response time indicates an average sound level taken over a period of time, perhaps as much as a second. Use a fast response time if the program material is very "active" and has lots of peaks, like loud drum beats. Slow response is perfectly suited for softer music or voice.

- Hold Button — A "hold" button lets you "freeze" the meter indication at any given time. You might use the hold button, for example, when holding the meter over your head, where you can't easily see the meter reading. Press the hold button, then read the meter.

- Peak Switch — A "peak switch" stores the loudest measurement taken by the meter during an interval. It is helpful if you want to determine SPL of the loudest portion of the program.

- Output Jack — An output jack may be provided to feed the signal to a recording device.

You will obtain more accurate results if you hold the meter at ear height of the *audience*. Hold the meter at a right angle to the sound source (not pointed at the sound source) as shown in *Figure 6-7*. Keep the meter as far away from your body as possible, so that it does not influence the reading in any way.

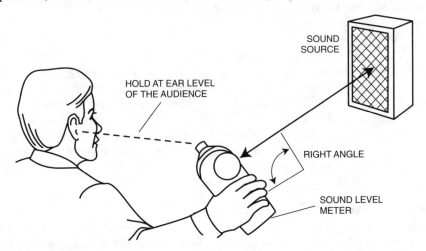

Figure 6-7. Hold the sound level meter so the microphone is at a right angle to the sound source. Place the meter at audience listening height. During the test, keep the meter as far from your body as possible.

The Effects of Sound Level With and Without and Audience

The sound check you perform before the actual presentation can only be an approximation, because the auditorium is devoid of people. As the room fills with people, the sound is absorbed, and even the tonal range of the sound is changed. You may find that the sound level between an empty and a full auditorium differs as much as 50 to 70 percent. For this reason, you will almost always need to increase the output sound level of the mixing console and/or the power amplifier when the room contains an audience. However, the mixing levels of the individual inputs seldom need further adjustment. Only experience will tell you how much an audience absorbs the sound in the sound system environment.

An audience can also influence the frequency response of the room. During the sound check, the sound may be bright and clear; but with an audience, it may be dull and flat. This is not uncommon. As the sound is absorbed by the people in the room, the highest frequencies are absorbed first. The sound then appears to be lacking in depth. You can compensate for this change in tonal quality if your sound system is equipped with an equalizer. Use the equalizer to boost the high frequencies; this will help restore the natural tonal balance of the sound.

OPERATING THE SOUND SYSTEM DURING A LIVE PERFORMANCE

With the sound check completed, the audience can be seated and the lecture or performance can begin. If the performance closely follows the sound check, leave the sound system equipment on, but move the master output slider (Master Fader) on the mixing console to minimum to prevent any signal from going through the system. Be sure not to move the input sliders that have been set for each input. Also turn off any battery-powered microphones to conserve their batteries.

If the sound check is performed much earlier in the day, you may turn the sound system off. This helps prevent accidental damage and tampering. To turn off the sound system, follow these steps:

1. Turn off the power amplifier.
2. Turn off the mixing console.
3. Turn off the preamplifiers in each microphone, if they are so equipped.

Before the performance begins — and preferably before the audience has arrived — turn on the sound system by following the steps given in "Testing for Sound Throughput" earlier in this chapter. Make a quick test to ensure that all microphones and other components are in good working order. Then while the audience enters the auditorium, switch off all mixer console microphone inputs (unless, of course, someone is using a microphone to address the audience) to keep the microphones from picking up unwanted noise. Then switch on the microphones again before the performance begins.

Making Minor Adjustments

It is inevitable that you will need to make final adjustments, called "tweaks" in sound engineer lingo, to the sound system during the actual performance. Such adjustments must be made slowly and carefully so that the audience is unaware of the change. An abrupt change is annoying, especially if you over-compensate and make the sound level too low or too high.

- If the overall sound level is too low, increase the output level of the mixing console and/or the power amplifier in small steps. Between each step, wait a moment and carefully listen to the sound. Avoid turning the sound level up too high, or acoustic feedback could occur.
- If the overall sound level is too high, you should also decrease it in steps, but you should do so as soon after the program has started as possible. The reason is that our hearing mechanism reduces its sensitivity somewhat when subjected to loud sound levels over time. If a level that is too high continues for even a short period of time, the audience may have difficulty hearing the program if you later decrease the level to a more appropriate level. The sooner you can adjust the sound level, the better.

Understanding Gain Staging

Your sound system is composed of a number of preamplifiers, intermediate amplifiers, and power amplifiers. Many of these amplifiers are equipped with their own level control. Most sound systems have three level controls (more elaborate sound systems may have additional level controls):

- Mixing console input level, for microphones and other input sources, such as tape decks or CD players
- Mixing console output level
- Power amplifier output level

Each level control works interactively with the others. It is important to properly adjust the level of each control in the correct sequence, usually beginning at the power amplifier output level and working toward the source inputs. Sound system novices often make the mistake of setting one level too low, then overcompensating by setting the other levels too high. This leads to increased noise in the system, as well as diminished sound quality.

Each level control represents what's known as a gain stage. As you adjust the level controls on the sound system, pay attention to the ratios between each gain stage. You obtain better sound quality and flexibility when the gain stage ratios are close to the same. For example, avoid the situation where the microphone inputs are set very low, and the mixing console output is set very high, as indicated in *Figure 6-8*. Noise and distortion are sure to occur with these settings. If there is more than a 3:1 difference between the gain stages — the microphone input is three times more or three times less than the mixing console output level, for example — adjust the levels to compensate for the disparity. Increase the microphone level and decrease the console output level. Of course, during a performance make this change slowly, so the audience is not aware of it.

Controlling Feedback

The bane of any sound system operator is acoustic feedback. The high-pitched squeal of feedback is annoying to the audience and makes the sound system operator look like a bumbling amateur. The remedies for feedback depend on physical factors as well as the capabilities of the sound system. As detailed in Chapters 3 and 4, one of the cardinal rules to reduce or eliminate feedback is to place the speaker systems in front of the microphones, and by using microphones with a directional pickup pattern.

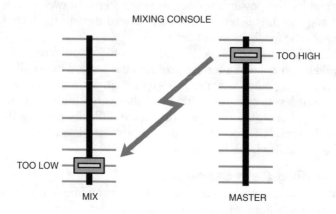

Figure 6-8. A mismatch gain staging occurs when one level control is adjusted too high or too low relative to another level control. Any deviance of more than about 3:1 could indicate poorly adjusted gain staging, and noise or other sound impairment could result.

Even with careful microphone and speaker system placement, some feedback may occur during a performance, especially if you are required to increase the sound level to compensate for a full auditorium of people. Depending on the available methods, you may wish to try some or all of the following:

- At the first sign of feedback, reduce the microphone levels until the ringing stops. In many cases, you must reduce a level to at least 50 percent of normal to get the feedback to stop. After the squealing has ceased, you can slowly increase each microphone level.

- If your mixing console is so equipped, turn on the feedback filtering switch. On most mixing consoles, this switch cuts out the frequencies most responsible for feedback. Because the tonal range of the sound system is affected, you should use the feedback filtering switch only if you really need it.

- Before the program begins, instruct your performers that if feedback occurs, they should not cover the microphone with their hands. This often makes the problem worse. Rather, they should allow you to control the feedback at your mixing console.

- If feedback continues, temporarily stop the program and move the microphone away from any speaker systems or boundaries that could be causing the sound to reverberate near the microphone.

A sound system that is prone to feedback is not installed and set up correctly. If you find feedback often occurs in the sound system, despite your best efforts to minimize it, carefully review the system layout, particularly the placement of microphones and speaker systems. If you cannot control their placement, consider replacing the microphones with models that have a more directional pick-up pattern.

Adding Special Effects

Some programs, especially musical performances, call for special effects. If your mixing console or power amplifier is so equipped, you might consider adding sound effects such as delay or echo. Use them sparingly; they can be intrusive to the audience. With many of the better mixing consoles, you can adjust the amount of delay or echo, as shown in *Figure 6-9,* in addition to switching these effects on and off.

A small amount of echo is often helpful for acoustic musical groups because it adds depth to the singing. Too much echo, however, can cause the voices and instruments to sound muddled and indistinct. A good rule of thumb: adjust the delay or echo so that it sounds good to you, then reduce it about 10 to 20 percent. Of course, prior to using these effects, you should obtain permission from the performers.

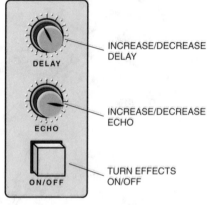

Figure 6-9. Some mixing consoles allow use of special sound effects such as delay and echo. The amount of delay and echo can be adjusted.

SPECIAL CONSIDERATIONS FOR BROADCAST OR RECORDING

We have been considering how to adjust the sound system controls to compensate for the effects of an enclosed room full of people. These requirements do not exist for broadcast or recording, since the sound is not heard in the same room as the performance.

You have two options if the program is to be broadcast or recorded: tap off the signal from the mixing console, or set up a completely separate mixing console just for the broadcast or recording. The latter is the preferred method, but in most instances, a separate mixing console and operator is not available due to cost. This means you will have to use the same mixing console for both the public performance and the broadcast or recording.

When a single mixing console is used, best results are obtained by simple compromise. You will need to monitor the program in progress both with your ears (for the public performance) and with fully-enclosed headphones (for the broadcast or recording). Adjust the mixing console and equalizer — if your sound system is so equipped — for best sound for both destinations.

Placing a microphone in front of a speaker system to obtain the sound for the broadcast or recording usually gives poor results. The microphone picks up a highly amplified version of the program, as well as any localized sounds from the audience. You will get far better results by sharing the output signal from the mixing console between the power amplifier and the broadcast or recording device. *Figure 6-10* shows a simple Y-connector used to split the output of the mixing console two ways. Use a Y-connector if your mixing console does not provide built-in multiple outputs.

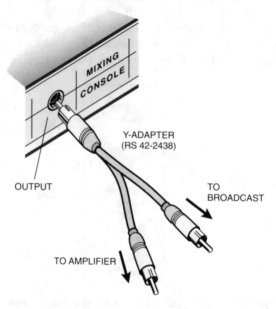

Figure 6-10. If the mixing console does not provide multiple outputs, use a Y-adapter to split the output to two cable lines. Then attach the ends of the adapter to the power amplifier and the broadcast link or recorder.

DIAGNOSING SOUND SYSTEM PROBLEMS

In the next chapter, you will learn about diagnosing problems within your sound system, including locating the source of internal hum, buzz, and other noise; finding faulty cables; and deciding whether you should repair or replace. You will also learn how to diagnose and fix poor sound quality, and how to track down and minimize external interference, such as static and electrical interference.

Maintaining and ⑦ Troubleshooting Sound Systems

INTRODUCTION

A sound system is typically a hardy beast. Unless it is abused, or used under adverse conditions; e.g., in wet or dusty conditions, it should give you many years of dependable service. To keep your sound system in tiptop shape, it's a good idea to routinely maintain it. Because most sound systems contain no moving parts, except for, perhaps, a cooling fan, routine maintenance is usually a simple matter of keeping the system clean. This chapter provides general cleaning procedures and recommends cleaning intervals.

Even a well maintained sound system breaks down from time to time. Many failures, such as a damaged cable or microphone, can be repaired easily. In this chapter you will find handy troubleshooting techniques to help you quickly locate trouble in your system. You will learn how to diagnose faults in the microphones, mixing console, power amplifier, and speaker systems. You will also learn how to find sources of hum, buzz, static, and distortion.

GENERAL CLEANING AND UPKEEP

They say that an ounce of prevention is worth a pound of cure. It's certainly true of sound system components. Keeping the microphones, mixing console, and amplifier clean goes a long way toward preventing more serious problems later. The following paragraphs describe general cleaning and upkeep of your sound system.

Checkup Schedule

How do you know if your sound system needs a preventive maintenance checkup? Experience is your best guide. As you become acquainted with your sound system, you will get to know when it needs a checkup and, possibly, a thorough cleaning. If the sound system looks dirty, it is; and you should clean it as soon as possible. A thorough cleaning and inspection is also called for if the system doesn't seem to be functioning up to par. For example, if the signal from a microphone is intermittent — sometimes the sound gets through and sometimes it doesn't — you should inspect the microphone and cable, and perhaps clean all connections.

The preventive maintenance (PM) interval for a sound system varies depending on many factors. If you use the sound system at least once or twice a week, you should give the system a PM checkup every few weeks. If your sound system is subjected to environmental extremes, such as heavy doses of dust, dirt, water spray (especially salt water), airborne oil, or sand, you may need to perform a PM checkup more frequently. Suggested PM intervals are given in *Figure 7-1* for systems that have light, medium, or heavy use.

Light Use

PM (Weeks)	3	6	9	12	15	18	21	24	27	30
Clean/dust exterior		•		•		•		•		•
Clean interior				•				•		
Clean connectors		•		•		•		•		•
Inspect cables			•			•				•
Inspect microphone battery				•				•		

Medium Use

PM (Weeks)	3	6	9	12	15	18	21	24	27	30
Clean/dust exterior		•		•		•		•		•
Clean interior			•			•			•	
Clean connectors		•		•		•		•		•
Inspect cables		•		•		•		•		•
Inspect microphone battery				•				•		

Heavy Use

PM(Weeks)	3	6	9	12	15	18	21	24	27	30
Clean/dust exterior	•	•	•	•	•	•	•	•	•	•
Clean interior		•		•		•		•		•
Clean connectors	•	•	•	•	•	•	•	•	•	•
Inspect cables	•	•	•	•	•	•	•	•	•	•
Inspect microphone battery			•			•			•	

Figure 7-1. Clean and inspect your sound system according to its use. Heavier use demands more frequent preventive maintenance. This is a suggested schedule; you may wish to set up your own.

CAUTION

Any time you work on equipment, remember to unplug the power cords before beginning — turning off the equipment power switch is not enough. Observe all safety requirements.

Unless you are an authorized repair technician for your sound system components, simply opening the equipment may void the warranty. Even if the warranty has expired, do not disassemble the equipment unless you are qualified to do so. Dangerous voltages and currents may be present inside the equipment.

User Manuals

An important measure in your efforts to minimize troubles with your sound system is to read the instruction manuals that came with your equipment. This may sound a bit obvious, but you might be surprised how many sound system operators and engineers never take the time to thoroughly read the user's manuals. The manufacturers of the equipment include the manuals so you can get the most out of your expensive purchase. Keep them at hand, and refer to them whenever you have a question.

General External Maintenance

At the top of your system PM list should be routine external cleaning. Each time wipe each component of the sound system with a dry cloth. Don't use dusting sprays; these actually attract dust, luring dirt back onto your equipment. Use a soft, sable painter's brush for those hard to reach places. Be sure to clean the ventilation slots as these are favorite hiding places for dust.

If you need to get rid of stubborn grime, apply a light spray of regular household cleaner onto a clean rag, then wipe with the rag. Never apply the spray directly onto a component, as the excess can run inside. Never apply a petroleum- or acetone-based solvent cleaner as it might remove paint and melt the exterior plastic parts. Also, some plastics, when in contact with a solvent, emit highly toxic fumes that could seriously injure you. Always read the directions and cautions on the labels of cleaning agents.

Cleaning Cords and Connectors

Inspect all system cables, wires, and connectors. Dirty or damaged cords and connectors can make your sound system "sneeze and wheeze." Repair or replace any cord that is damaged. If you suspect that a cable is bad, try a replacement to see if the problem is solved.

You can test the cable for continuity or for shorted conductors by using a volt-ohm meter as shown in *Figure 7-2.* Attach (or just touch) the test probes of the meter to the pins at each end of the cable for a particular path. If the wire is not broken, the meter should read zero ohms. Repeat for each wire in the connector. If a cable has a grounded shield, you can check the continuity of the shield by placing the meter probes on the outside metal shell at each end.

Check for shorts to adjacent pins and to the connector shell by applying the test leads to these points (be sure not to hold onto the metal portion of the probes — you might get false readings). The meter should read an open circuit, which is indicated in *Figure 7-2b* as an overrange reading on the display.

When you're satisfied that all cables and wires are in good shape, clean all the connectors to ensure good electrical contact. Suitable contact cleaner is available at Radio Shack stores in aerosol spray cans or wipe-on pen form, as shown in *Figure 7-3.* You can also use a cleaner designed for magnetic heads in video cassette recorders. Apply the cleaner to the connector with a cleaning pad, shown in *Figure 7-3b,* or cotton swab.

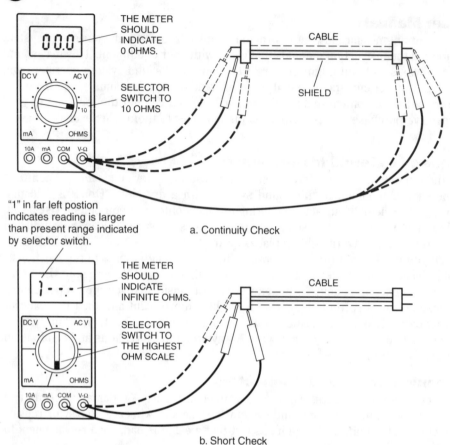

THE METER SHOULD INDICATE 0 OHMS.

SELECTOR SWITCH TO 10 OHMS

CABLE

SHIELD

"1" in far left postion indicates reading is larger than present range indicated by selector switch.

a. Continuity Check

THE METER SHOULD INDICATE INFINITE OHMS.

SELECTOR SWITCH TO THE HIGHEST OHM SCALE

CABLE

b. Short Check

Figure 7-2. Use a volt-ohm meter to test the continuity of all conductors in cables and to check for shorts. Set the meter to read ohms.

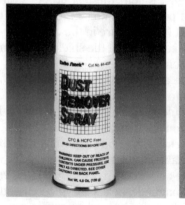

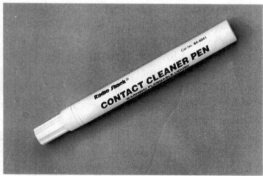

a. Aerosol Spray Cleaner   b. Pen Contact Cleaner

Figure 7-3. Cleaner for electronic equipment and contacts is available in aerosol spray or brush-on pen form. The pen cleaner is more efficient, but is not as useful for getting into hard to reach cracks and crevices. Cleaning pads are available to wipe on cleaner. (Courtesy of Radio Shack.)

Interior Cleaning

Sound equipment used in extremely dusty areas, or left unattended for an extended time, often needs cleaning on the inside. This cleaning should be done only if you feel competent in removing the protective covers of the equipment, and only if you observe all safety precautions. *Remember: you may not want to disassemble the equipment if doing so will void the warranty. Check the equipment warranties carefully.*

There are three ways to remove dust and dirt — brushing, vacuuming, and blowing. After removing the outer cabinet of the equipment, use a soft brush to wipe away excess dust and dirt and/or a small hobby vacuum cleaner to remove sediment. Or the dust and dirt may be blown out. You can purchase a small can of compressed air at a photographic shop for about $3. Squirt it liberally inside the equipment, keeping the can upright to prevent dripping of the propellant. Position the air flow so that debris is blown out of the equipment, not deeper inside.

Avoid using household cleaning sprays, as these not only can leave a residue that could impair proper operation, but also are water-based, and might cause a short circuit. *DO NOT use a cleaner that contains a solvent of any kind. DO NOT clean your sound system with an oil-less lubricant such as WD-40.*

Nicotine sediments and caked-on dirt may need a more rugged cleaner recommended for application on mechanical and electronic equipment. These cleaners have a built-in degreaser that attaches the oil build up; these are fine for use in your sound system equipment, as long as they are designed for direct application onto electronic parts. Whatever cleaner you use, make sure it leaves no residue after drying (which usually takes less than 15 seconds). If the cleaner leaves a noticeable residue, it is unsuitable for use inside your sound system components.

Servicing Microphones

Microphones used in sound systems generally require little attention, apart from periodic dusting. However, most quality sound system microphones use a built-in amplifier, which is powered by one or more batteries installed in the microphone. Periodically check the battery in each microphone to make sure it is not leaking and test the battery condition to make sure it has sufficient voltage under load.

Battery electrolyte is generally dark brown in color, but the corrosion effects of the electrolyte can cause a white powdery substance to accumulate around the battery. Be on the lookout for this, and replace the battery if you see signs of leakage or bulging. If battery leakage has contaminated the battery terminals inside the microphone, you can clean the terminals with a contact cleaner. Use a pencil eraser for hard to remove contamination.

Weak or worn out batteries should be promptly replaced, as these are the most prone to leaking. You also don't want to be caught with a dead microphone battery during a performance. Test the condition of the battery using a battery tester. Do not use an ordinary volt-ohm meter, as the meter does not test the battery under load. Battery testers, such as the models shown in *Figure 7-4,* test the battery under load, and the results are much more accurate.

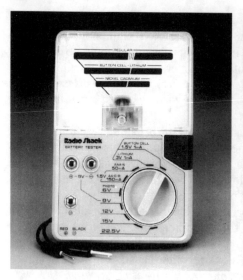

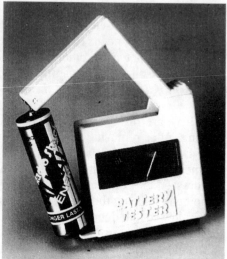

a. 8-Range   b. Pocket Size

Figure 7-4. Use a battery tester to test the condition of microphone batteries. The battery tester tests the battery under load, therefore, it yields more accurate results than just measuring the voltage with a volt-ohm meter. (Courtesy of Radio Shack.)

TESTING OPERATION OF SOUND SYSTEMS

After you've cleaned your sound system, take a little time to test its operation. Follow the procedure in Chapter 6 for powering up the system. Test all microphones and other inputs. You may wish to keep a log book of all the maintenance checks and cleaning you've performed on your sound system. A sample log is provided at the end of this chapter. In your log, be sure to note the exact checks and maintenance procedures you performed, and when you performed them.

TROUBLESHOOTING PROBLEMS WITH SOUND SYSTEM COMPONENTS

Despite the regular preventive maintenance, problems may beset your sound system, causing either poor sound or no sound. There's no need to panic. In many cases, the real problem is minor and can be repaired in a matter of minutes if you follow set procedure.

The troubleshooting procedure depends on how complicated the system, what resources you have available, and the fact that you cover all the bases. Be sure to investigate all possibilities. Don't take anything for granted. For example, a very common problem is that someone has pulled loose a power plug. Don't assume a piece of sound equipment is plugged in and functioning. Verify that it is by checking the power indicator. You might be surprised how many problems with sound systems are merely the result of components not plugged in or turned on!

If a sound system suddenly goes out, it is usually caused by a bad cable or a bad connector. Either a conductor has broken or a connection becomes bad because of a dirty, corroded, bad connector. The first and easy technique is to have a cable to substitute for a suspected cable. It is particularly handy in situations when you don't have time to test every detail. Because sound systems are composed of many components, whenever possible, tackle problems by substituting a new cable or a new component. Or bypass a suspected cable or component so that it is eliminated from the system.

As an example, consider the most common problem in sound systems — no output sound when someone speaks into a microphone. Here is a brief trouble-shooting procedure:

1. Make sure the microphones are turned on. Make sure all equipment is plugged in and turned on. Make sure all input and output controls are adjusted correctly. Make sure all interconnecting cables are in place and connected.

2. If all microphones are affected, then try using a tape or CD input to see if you get an output from the speaker systems. If so, then the problem is in the micro-phones or microphone inputs to the mixing console. If you don't get sound with a tape or CD input, then the problem is in the mixing console, power amplifier, speaker systems, or an interconnecting cable. Bypass the mixing console and connect directly into the power amplifier (if possible). Replace the interconnect-ing cables one at a time.

3. If only one microphone is affected, then, most likely, that microphone, its bat-tery, or its connecting cable is defective. It's also possible that the input circuit in the mixing console is bad. Plug in a different microphone and cable into that mixing console input and see what happens.

Once you've replaced or bypassed the bad faulty component, you've isolated the problem and can focus on fixing it.

The troubleshooting charts of *Table 7-1, 7-2,* and *7-3* that follow will help you troubleshoot the most common maladies of the major elements of the sound system: microphones, mixing consoles, amplifiers, and speaker systems. To use a chart, just locate the problem in the left column, and read the possible causes and solutions in the right column.

Table 7-1. Problems with Microphones

1. No sound from microphone.	• If microphone has on/off switch, check that the switch is in the ON position. • Check for bad connection from microphone to mixing console or amplifier. • Check voltage of battery in the microphone (if so equipped). Battery should also be inserted properly, and the battery terminals must be clean.
2. "Crackling" sound from microphone.	• Check connection from microphone to mixing console or amplifier. Connection may be loose or dirty. • Check battery. • If cracking is caused by wind, shield microphone element from wind (use a windscreen if necessary). • If cracking is caused by electrical interference, move away from the device (such as an electric motor) causing the interference.
3. "Buzzing" sound from microphone.	• Microphone not properly grounded. Check connector.
4. Microphone causes electrical shock when you touch it.	• Microphone and/or entire sound system not properly grounded. TURN OFF IMMEDIATELY and check for proper ground. Use tests outlined in Chapter 5.

Table 7-1. Problems with Microphones (cont.)

5. Microphone causes acoustic feedback.	• Move microphone so that it is not in direct line with a loudspeaker. • Reduce microphone gain. • Turn on feedback control circuit on mixing console (if so equipped).

Table 7-2. Problems with Mixing Consoles

1. No power.	• Mixing console switched off or power cord is unplugged. • Check fuse (if so equipped). Replace fuse if needed. • Check power cord. Repair if bad.
2. No signal on one or more inputs.	• Check connections to mixing console. • Verify input switches (if so equipped) are in proper position; e.g. "Mic" for microphone inputs or "Line" for line inputs. • Verify that gain controls are at least at the one-third position.
3. Sound is distorted.	• Verify input switches (if so equipped) are in proper position; e.g. "Mic" for microphone inputs or "Line" for line inputs. • Check gain settings. Master gain should not be excessively higher than gain controls for individual inputs.
4. Sound contains excessive noise.	• Determine type of noise and remedy accordingly: hum is typically caused by ac induction; buzz and static is typically caused by electrical interference such as a faulty fluorescent light; hiss is typically caused by improper setting of gain controls.

Table 7-3. Problems with Power Amplifiers and Speakers

1. No power.	• Mixing console switched off or power cord is unplugged. • Check fuse (if so equipped). Replace fuse if needed. • Check power cord. Repair if bad.
2. No sound or low sound volume.	• Verify proper input from microphone or mixing console. • Check speaker switches (if so equipped). Be sure "A" speakers are switched ON if speakers are connected to the "A" terminals. • Check gain controls. Slowly increase until you hear the sound. • Check speaker connections on back of amplifier.
3. Sound from one speaker is distorted.	• Power output too high for speaker; reduce output gain. • Check for physical damage to the speaker, especially to the speaker cone. Replace speaker if necessary. • Inspect for foreign material (fiberglass batting, pieces of wood) in speaker enclosure; remove and repair as necessary. • Check speaker connections on back of amplifier. Make sure that speaker system is connected between the right "ohm" terminal and ground (common).

SHOULD YOUR REPAIR OR REPLACE?

Part of troubleshooting and maintaining a sound system is deciding when to keep using what you have, and when to "throw in the towel" and replace a faulty component with a new one. Deciding whether to repair or replace is not always an easy task, as you must take into consideration the cost of repair if you are not doing the repair yourself, the cost of a suitable replacement, and any extra cost associated with adding a new component to an existing sound system.

If you are able to repair the equipment yourself, you may wish to do so rather than buy a new component, but do take into consideration the value or your time — especially if you work for an employer. Suppose it takes you two days to fix a piece of sound system gear. How much will that cost your employer. If it's about the same or more than a replacement, you should consider replacing the faulty component.

Deciding on whether to repair or replace is often easier if you are not doing the repair yourself. You can more easily judge which is the better value: taking the unit in for repair, or replacing the unit outright with a new model. The choice is obvious if the repair costs $100, but the replacement costs $500. On the other hand, there is generally little to gain in spending $50 to repair a microphone that you can replace for $40.

However, beware of judging the value of repair versus replacement on cost alone. Also consider the general condition of your gear. If you are constantly sending in one component or another for repair, you will probably save money in the long run by purchasing a replacement. If a piece of equipment costs more to maintain in a one- or two-year period than a replacement, it may be worth your while to replace it with a newer, more dependable model. Keep a record of the repair bills for all your equipment so you know which components are draining the maintenance budget.

DIAGNOSING POOR SOUND QUALITY

The only thing worse than a sound system that doesn't work is a sound system that delivers poor sound quality. In previous chapters, you read how to improve the quality of sound from a sound system by observing proper microphone and speaker placement, creative use of sound absorbing materials, and correct operation of the mixing console gain controls. In this section you'll learn about correcting the problems of sound caused by bad or loose connections, or nearby electrical fields that are interfering with the sound system. These faults can cause annoying hum, static, hiss, and distortion.

As a general rule of thumb, the type of noise often signifies its origin; therefore, you can use the noise to help find its source. While this method is not 100% guaranteed to lead you to the guilty party, it will get you started in tracking down the source of the problem so that you can remedy it.

Hum

A low-frequency hum is one of the most common maladies of sound systems. The hum you hear is the 60-Hz alternating current used to power the sound system. The volume of the hum can be soft or loud, depending on its source, and where the noise enters your sound system. By far, the most common cause of hum is poor grounding of the sound system components. Proper ground is required not only for safety reasons, but also to assure proper ground levels between all components.

One source of hum is the microphone. Hum can be induced and amplified when the microphone uses a metal casing, and this casing is connected to the ground lead of the microphone input. The loudness of the hum changes when a person touches the microphone because the person's body picks up signal to add to the input. Often, the hum is softer when the microphone is held, as the person's body acts to ground the microphone. In any case, there is variation in the hum. Under normal circumstances, the microphones in your sound system should not be creating a hum. If they are, look for reversed connections (signal and ground) of the microphone inputs. You may also wish to try higher quality microphones, such as those that use the XLR type connector.

Ground Loops or Improper Ground

Hum can also be induced when the components in the sound system do not share the same electrical ground. This is common when the components are not plugged into the same outlet. Outlets separated by a distance may have different ground potentials. This is demonstrated in *Figure 7-5*. When the voltage to ground is measured at outlet A, it is 0.5V. When the voltage to ground is measured at outlet B, it is 3.7V. There is a difference of 3.2V between the two outlets, which can set up a ground loop condition that can cause hum.

Hum caused by improper ground can also occur if the grounding prong of the ac plug for the sound system component has been bypassed or removed. Whenever possible, plug all components into the same outlet, or at least in outlets on the same electrical circuit. And be sure the components are properly grounded. Following these procedures will help reduce the hum induced in your sound system.

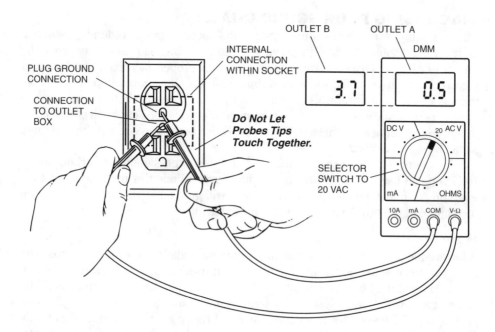

Figure 7-5. The actual voltage of the ground conductor in an electrical socket can vary between outlets. For example, at one outlet the actual voltage at the ground connection, relative to earth ground, may be 0.5 volt. At another outlet the voltage may be 3.7 volts. This difference in ground potential may produce hum when sound system components are plugged into different outlets.

Transformer Action

Another source of hum is caused by induction through ac cords and signal cables, especially microphone cables. If an ac cord is positioned too close to a signal cable, the 60-Hz current can be inductively coupled into the signal cable. The basic principle is the same as a power transformer — an ac current through one set of wires is inductively coupled to another set of wires. Hum caused by induction is typically caused by placing loops of ac cord and signal cable too close to one another as shown in *Figure 7-6a*. The loops act as coils inside a transformer, and under the right conditions, a great deal of hum can result. Eliminate hum caused by induction by not coiling ac cords and signal cables. Keep the ac cords and signal cables physically separated. If ac cord and signal cable must cross paths, try to have them cross at right angles, as shown in *Figure 7-6b*. Avoid routing them in parallel. You can also use pieces of carpet or blankets to physically separate the ac cords from the signal cables.

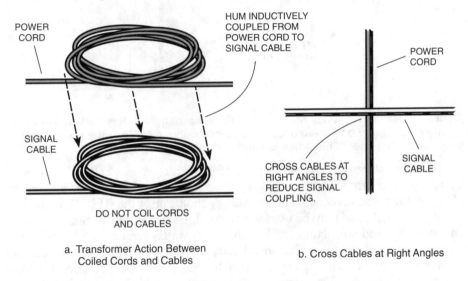

a. Transformer Action Between Coiled Cords and Cables

b. Cross Cables at Right Angles

Figure 7-6. Hum can be inductively coupled from an ac power cord to a microphone or other signal cable. Placing ac cords and signal cables at right angles to each other minimizes the effects of inductive pickup. Avoid looping the power cord and signal cable and placing them close to one another.

Buzz and Static

A buzz is characterized as a constant noise that has a fairly steady frequency. The frequency of the buzz depends on its source; it might be low-pitched or high-pitched. Conversely, static is characterized as a noise that occurs randomly and makes a popping or crackling sound. Buzz and static are often caused by the same thing — radio frequency interference (RFI). There are many types of RFI and their results in the sound output of the sound system may be quite different. One common source of RFI is an electrical appliance (which may be malfunctioning) that is operating close to one or more components of the sound system.

Buzz

A steady buzz is often caused by radio frequency emissions from a ballast in a fluorescent light fixture. The RFI is carried through the air, as indicated in *Figure 7-7,* the same way that an AM radio station broadcasts to nearby radios. The RFI is picked up by the sound system components, especially those that lack sufficient shielding, such as low quality microphones or cables. The RFI of newer style ballasts has been reduced, but older models, as well as defective ballasts, can generate prodigious amounts of electrical noise.

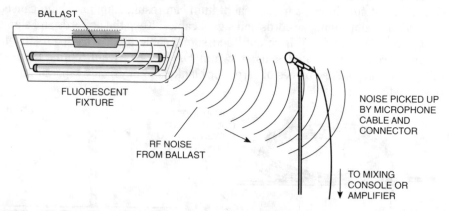

Figure 7-7. Radio frequency noise from a fluorescent light fixture ballast can travel through the air and be picked up through a microphone cable or the connectors in the microphone itself. This noise is amplified as an annoying buzz.

Dimmer switches are another source of RFI which causes buzz. The dimmer uses a silicon-controlled rectifier (SCR), which switches power to the lamp on and off many times each second. The switching action causes the RFI. Dimmers can be large or small, and may be used where you least expect them. Small dimmers are typically used with lamps of less than 300 watts, such as the lights over a podium. Larger dimmers, which are often used to dim lamps in a school theater or auditorium, may be controlled by a remote lighting panel. In operation, this panel can produce a tremendous amount of RFI. If you sound system or microphone is placed anywhere near this panel, buzzing is sure to result. Turn lights on and off to detect the problem.

Though less common, buzz can also be caused by electrical noise traveling through the ac wiring. Such noise is typically caused by appliances, such as refrigerator or air conditioning compressors. The noise is worse if the compressor is old. Modern compressors are much more efficient and generate far less RFI than previous models. The best way to combat noise caused by electrical equipment is to dampen it at the source. This is typically done by attaching a filter to the appliance. The filter plugs into the ac outlet; the appliance then plugs into the filter.

Filters for heavy duty appliances can be expensive because they must be rated for high wattage. A cheaper alternative is to use an RFI filter on your sound system. The filter plugs between the sound system components and the ac outlet. Such filters are typically designed to reduce TV interference, and may be found in the television section of home improvement centers.

Static

Whereas buzzing is a constant stream of noise, static is aperiodic (random crackling) and its sources are harder to track down. The ignition system of a car, especially an old car, is a typical source of static. The static is generated by the high-voltage coil and the spark discharge within the engine. This spike of RF noise is then carried through the air, where it can cause snaps, crackles, and pops in your sound system. The best remedy is to suppress the noise at the source. This can be accomplished by replacing the spark plugs with resistor type plugs, and replacing old or defective spark plug wires, distributor components, or coils. However, since the owner or operator of such equipment may not be cooperative, you may have to take action at your end.

Random single pops of static noise can be caused by dirty connections in your sound system connectors, especially if the cables are moved around while the system is "live."

If it is not possible to repair the source of the static, or the source of the static cannot be found, you should consider improving the shielding for your sound system. Most static is picked up by the long lengths of signal wires between microphone and mixing console, and the mixing console and the power amplifier. For ultimate rejection of static, try XLR type microphones with a balanced input. The balanced input helps to reject random noise such as static, even in relatively long lengths of cable. In addition, the XLR connector also has the capability to use a second shielding connector as shown in the illustrations in Chapter 5.

Distortion

Sound is distorted when the shape of the output signal differs from the input signal. The typical distorted output has sound signals that are "clipped" — the louder portions of the output signal have flat tops. The result of even modest distortion can be sound that is barely intelligible.

There are many types of distortion, and many causes for it. One common cause is to crank up the output level control on the amplifier so the amplifier is pushed beyond its design limits, trying to pump out more watts than it can deliver. Reducing the output level of the amplifier restores the quality of the sound. Distortion can also occur when using the wrong inputs to the mixing console or power amplifier, and it's this kind of distortion we're most interested in, because it is the most common.

Revisiting Impedance Matching

As you know from Chapters 2 and 3, microphones are designed with a certain output impedance, and they expect a particular impedance at the input connection. We've stressed in these chapters that, except in the most demanding applications, the exact numerical impedances of the microphone and input are not important. They may be classified simply as low impedance or high impedance. But proper impedance matching is very important. It is necessary to connect a low-impedance microphone to a low-impedance input; or a high-impedance microphone to a high-impedance input. Considerable distortion can occur if you connect a high-impedance microphone to a low-impedance input, or vice versa.

Imbalance in Levels

Distortion can also be caused by an imbalance in levels. All input devices — from microphones to tape players to CD players — generate a voltage output. The actual voltage is usually fairly small, often one volt or less. Your mixing console is designed to accept inputs at a given level. Inputs are rated by their typical use, and it's important to not mix inputs, or else distortion could result. Following are typical levels for microphone, tape, phono, and CD inputs:

Table 7-1. Typical Voltage Levels of Sound System Inputs

Input Type	Typical Level
Microphone	7 mV
Phono, magnetic	20 mV
Phono, ceramic	1.2V
Line-level (CD, tape)	1V

As you can see, an input level mismatch will occur if you connect a 1V line-input source such as a CD player to a microphone input, which is designed for a device with an output of just 7 mV (millivolts, or thousandths of a volt). The CD signal would overdrive the amplifier setup for the microphone input and cause severe distortion.

Some mixing consoles are equipped with a single input and a function switch for selecting the input type: microphone, phono (usually magnetic), or line. Be sure to set the switch for the proper input. The setting of this switch is commonly overlooked; you should get into the habit of checking the position of the input switch each time you use the sound system.

Notice the differences in output levels from a magnetic and ceramic pickup from a turntable arm. Turntable arms that use a ceramic cartridge have a much higher output voltage, and should be connected only to a ceramic phono input on the mixing console. Turntable arms with magnetic cartridges (the most common type manufactured today) have a relatively low output voltage, and should be connected to the magnetic phono input.

While the output levels of a microphone and magnetic pickup are very similar, you should not connect a microphone to a magnetic phono input, or vice versa. One reason is the differences in impedance. The magnetic phono input is designed for a fairly high impedance of about 50k ohms, whereas a low-impedance microphone input is designed for less than 2k ohms. Another reason, as we pointed out in Chapter 3, is that a RIAA equalization circuit is used in the phono input. Without the equalization the audio from the turntable will lack bass tones, and will sound unnaturally flat, or even tinny.

Levels are also a consideration between the mixing console and the power amplifier. The output of most mixing consoles is a line-level output of from 1 to 2 volts; you should connect this output to the line-level input of the power amplifier. DO NOT connect the output of the mixing console to a microphone input or other input on the power amplifier; extreme distortion could result.

Sound System Maintenance Journal

A sample of a sound system maintenance journal is in the Appendix. It should be used to record anything that you do to your sound systems.

Installing and Using Sound System Add-Ons

INTRODUCTION

The *basic* sound system is made up of a microphone, mixing console, amplifier, and speaker as shown in *Figure 3-1;* however, many sound systems also include a wide variety of supplementary equipment.

This chapter introduces you to some of the equipment you can add to your sound system to improve its performance or make its use more convenient. Equipment such as wireless microphones, monitoring headphones, assisted hearing systems, CD players, tape decks, turntables, and special effects such as reverberation (reverb).

WIRELESS MICROPHONES

The bane of any sound system operator is microphone cables. Invariably, the cables get broken, crushed, knotted, and tripped over. Some performers, including musicians, singers, and lecturers, are greatly hampered by the microphone cord. They prefer to walk around the stage unimpeded by the tangle of a microphone cable. Microphone cables, even the best ones, may be a primary source of interference. Placing them too close to ac cords or electrical outlets can cause excessive hum and static.

Wireless microphones are just as the name implies — microphones without cables. The wireless version uses a one-way radio signal (often in the FM broadcast band of 88 to 108 MHz) to transmit the sound from the microphone pickup to a receiver placed near the mixing console. The receiver output plugs into a *Line In* jack on the mixer.

There are a number of styles of wireless microphones and prices vary from under $50 to well over $500. The more expensive models are intended for professional theatrical use. The basic wireless microphone is shown in *Figure 8-1*. It is designed to be held or mounted on a stand. The unit consists of two parts: one part houses the microphone element, a low-power RF transmitter, and an antenna; the second part is the receiver. The microphone is powered by batteries. It is extremely important to regularly check, and if necessary, change these batteries to prevent microphone failure during a performance. In most wireless microphone models, the receiver is powered by an ac outlet, but may also be powered by batteries. The receiver uses a telescoping antenna to pick up the signal from the wireless microphone transmitter.

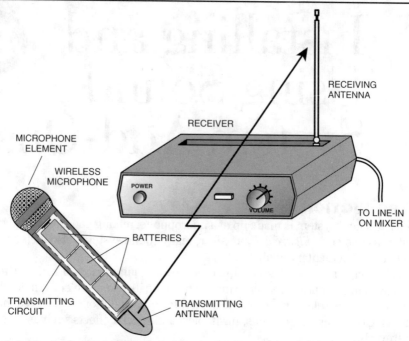

Figure 8-1. The typical hand-held wireless microphone operates in the FM broadcast band (80-108 MHz) or 170 MHz radio frequency range. In the better models, the microphone antenna is hidden inside the housing.

Another style of wireless microphone uses a lapel pickup with a thin, short cable connecting it to the transmitter as shown in *Figure 8-2.* The pickup attaches to the user's clothing as close to the mouth as practical. The transmitter attaches to any convenient part of the user's clothing, usually a pocket or belt. The antenna is usually a short flexible wire that hangs down from the transmitter housing. This type of wireless microphone is preferred by lecturers and some musical performers because it is not necessary to hold the microphone.

A variation of this style has a small boom microphone mounted on a headset which the performers wear on their heads. The microphone can be adjusted to be very close to the performer's mouth for maximum pickup and maximum exclusion of other sounds. This style often has headphones to provide monitor sound to the performer.

The RF power output of wireless microphones is relatively low for two reasons. First, the Federal Communications Commission (FCC) limits the power output to prevent interference with nearby radios, television sets and radio transceivers. And second, the lower power helps ensure that the microphone signal remains in the building rather than being broadcast so that people in the neighborhood can pick up the signal on their FM radio or scanner!

Because a wireless microphone's transmitter has low output power, the receiver should not be placed more than 100 to 200 feet from the microphone to ensure reliable and high quality reception. If the mixing console is located in a sound booth, you may find it necessary to locate the receiver outside the booth, as shown in *Figure 8-3,* so there is no structure between it and the microphone on the stage.

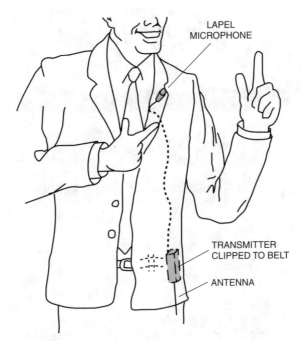

LAPEL
MICROPHONE

TRANSMITTER
CLIPPED TO BELT

ANTENNA

Figure 8-2. A lapel wireless microphone consists of a miniature microphone pickup and a transmitter unit which attaches to a pocket or belt. A thin cable connects the pickup to the transmitter unit.

Many professional-level wireless microphones operate at 170 MHz, which is above the FM broadcast band, because the higher frequency is not as prone to interference. As of this writing, the FCC requires a permit to use wireless microphones that operate in the 170 MHz range. You don't need to pass a test to obtain the permit. Rather, you pay a permit fee and sign an agreement where you promise to use the equipment in accordance with FCC regulations.

MICROPHONE ACCESSORIES

While the microphone may be the primary input source for your sound system, the microphone is considered incomplete without a few accessories. First and foremost is a stand. The stand holds the microphone so the performer or speaker doesn't have to. Microphone stands are often used by musicians who must have both hands free to play their instrument. Singers who are not playing an instrument often prefer to hold the microphone because it gives them something to do with their hands.

Figures 6-1 and *6-2* show some of the types of microphone stands. A floor stand with adjustable height allows the microphone to be positioned in front of a performer who is sitting or standing. Attach a microphone boom to a floor stand to position a microphone close to an instrument, such as an acoustic guitar or piano, while it is being played. Table (or desk) stands hold a microphone placed on a table or podium. For podium use, a "goose neck" flexible attachment permits the speaker to easily adjust the position of the microphone.

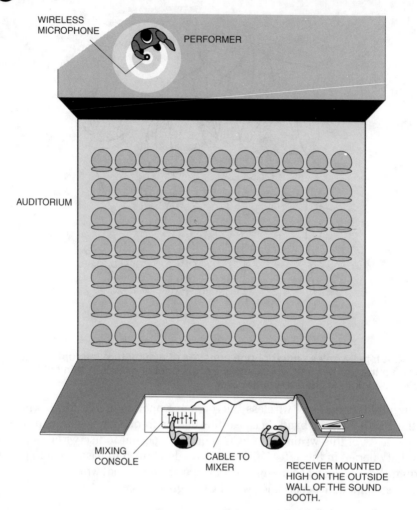

WIRELESS MICROPHONE

PERFORMER

AUDITORIUM

MIXING CONSOLE

CABLE TO MIXER

RECEIVER MOUNTED HIGH ON THE OUTSIDE WALL OF THE SOUND BOOTH.

Figure 8-3. For mixing consoles located inside a sound booth, locate the wireless microphone receiver outside the booth, in direct unobstructed line to the microphone on the stage. For best reception, position the receiver high on the wall.

A windscreen is a piece of acoustically-transparent material that helps prevent gusts of wind from entering the microphone. Windscreens are typically used outdoors, where wind can cause a "booming" sound when it hits the element of the microphone. Many speakers and singers also prefer to use windscreens on their microphones for indoor use as well. They feel it gives a clearer, more distinct sound.

Windscreens (sometimes also called windsocks) are seemingly simple devices, often just a hollow half-ball of special foam which you simply slip over the microphone. However, windscreens are not simply pieces of regular cloth or foam like you can buy at the general store. To be effective, the windscreen must pass sound waves, but block most of the wind, which is a tall order, since both are movements of air. Windscreens are designed to allow the maximum passage of sound waves, yet hold back undesirable wind. For this reason, you should not substitute ordinary cloth or foam for a windscreen.

MONITOR HEADPHONES

As a sound system operator, you must be certain during a performance that the sound from each input is at its proper level. For example, careless adjustment of the sound system controls can let the musical accompaniment drown out a singer. Depending on where you are located in the performance hall, you cannot always trust your "naked" ears as you listen to the sound through speakers.

More than likely, you will be located either very close to the stage or at the very back of the auditorium. Or, you may be situated off to one side of the auditorium. None of these locations are ideal listening posts. If you are positioned close to the stage, for instance, your ears may be unduly influenced by the direct (unamplified) sound from the stage. If you are positioned close to a wall (the side or back of an auditorium), your ears may be unduly influenced by reflected sound.

This is where monitor headphones are useful. Monitor headphones fully enclose your ears, allowing very little sound from the outside to get in. You hear mostly the sound from the headphones. You plug the headphones into a headphone jack on the mixing console or power amplifier. In this way, you can hear the sound relatively unobscured, and can make better judgments on how the various input levels should be set. Standard stereo headphones with full, soft cushions that totally enclose your ears are adequate for this job. Don't use "lightweight" or open-air headphones that don't block outside sound.

NOTE

Be sure that you DO NOT use the monitor headphones to adjust the master gain (volume) of the sound system. Use your ears — or better yet a sound level meter — for this job. The sound level through the monitor headphones is not at all guaranteed to be the same as the sound from the speakers.

ASSISTED HEARING SYSTEMS

Not everyone in your audience may have perfect hearing. You may want to offer assisted hearing systems for those who are hearing impaired. Assisted hearing systems come in many different forms. The model shown in *Figure 8-4* is a self-contained headset with stereo microphones, amplifier, stereo headphones, 3-band equalizer, balance control, and volume control. Many hearing impaired people are also sensitive to certain sound frequencies. A unit with equalization adjustments helps the users tailor the sound to suit their tastes. Another type of assisted hearing system is similar to some old-style hearing aids. It has a microphone and amplifier with volume control in a small box that the listener can hold or clip to clothing. The lightweight headphones connect to the box with a thin cable.

Another type of assisted hearing uses wireless technology, either radio frequency (typically using the FM radio band) or invisible infrared light. These devices tend to be more expensive, but they are also more effective. A transmitter is situated above the audience and broadcasts its signal so that it covers the entire room. Listeners in the audience wear the receiving headphones.

8

Figure 8-4. Amplified listeners provide individual amplified sound to members of the audience who are hearing impaired. (Courtesy of Radio Shack.)

Most wireless headphones consist of the base transmitter station and the receiving headphones. You can often buy extra headphones — typically on a special order basis — so that you can use one base transmitter station for multiple listeners.

SOUND INPUT SOURCES

While the microphones represents the most common input source in a sound system, most sound systems also rely on alternative sources, including CD players, cassette tape decks, and turntables. For most amateur and semi-professional sound system setups, it is generally acceptable to use CD players, tape decks, and turntables designed for home use (however, check local electrical codes before purchasing). On the other hand, if you're in charge of a professional sound system, you probably want the more rugged professional versions. In addition to being built for public use, the pro versions typically offer more control which allows you to tailor the device to your sound system.

CD players, tape decks, and turntables are designed for easy connection to your sound system. Assuming you are using a mixing console, each device connects to the console with a shielded cable with RCA-type phono connectors. The cable should be as short as possible to reduce the chance of picking up interference. Keep the following in mind when connecting alternative sound sources to the sound mixer:

- If the mixer has separate "MIC" and "LINE" inputs, be sure to use the "LINE" input for CD players and tape decks. Many mixers have one "universal" input and provide a switch for microphone or line use.
- Do not connect a turntable to any input marked "MIC" or "LINE." Instead, use only the input specially marked for phono use. The phono input is designed to accept the typical signal level from a turntable equipped with a magnetic cartridge. The input also has special equalization that helps restore the recorded sound to its proper tonal balance.

- If you use more than one alternative input source — such as a CD player and tape deck — but have only one line-level input, use an outboard mixer to "pre-mix" the sources. An outboard mixer is usually equipped with four or five line-level inputs and some have inputs for microphones; you can adjust the level for each input. The output of the outboard mixer then connects to the one line-level input of your master mixing console. *Figure 8-5* shows a basic outboard mixer which has four line-level inputs and one microphone input.

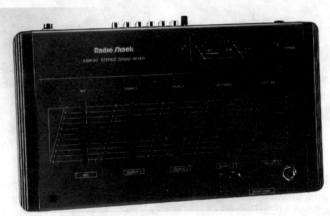

Figure 8-5. Use an outboard mixer to provide additional inputs when using multiple sound sources, such as a CD player, tape deck, and microphone. The output of the outboard mixer connects to a line-level input of the master mixing console. (Courtesy of Radio Shack.)

SPECIAL EFFECTS

Many of the better mixing consoles are equipped with a built in reverb/echo feature. Use this feature to add "depth" to the sound. Echo and reverb, when used correctly, can turn ho-hum singers into first-class cabaret performers (well, not really, but every little bit helps!). Echo and reverb can also be used to add special effects — for a spooky Halloween presentation, for example.

Don't despair if your mixing console lacks reverb and echo capability. You can often add it with an external special effects mixer. The mixer can be used in either of two ways:

1. Place the special effects mixer between a microphone and mixing console.
2. Place the special effects mixer between the mixing console and the power amplifier. Adjust the amount of reverb/echo for the entire sound processed by the mixing console.

Some mixing consoles are equipped with a "EQ" or special-effects output. This allows you to use an outboard equalizer or special effects mixer as if it were a part of the mixing console. The advantage of this method is that the mixing console can remain directly connected to the power amplifier. In this way, the output level controls of the mixing console still do their job of adjusting the final sound level.

If you're looking for way-out effects, try a special effects generator, such as the model in *Figure 8-6.* This generator creates dozens of digitally-sampled sounds, including cat meows, dog barks, cow moos, and much more! The generator can be programmed to create a sequence of special sounds.

Figure 8-6. Use a special effects generator to add unique sounds to your presentations. (Courtesy of Radio Shack.)

Figure 8-7. An all-in-one portable public address amplifier can be used for small-scale sound systems and as a monitor speaker for performers. (Courtesy of Radio Shack.)

ALL-IN-ONE PORTABLE PA SYSTEM

Not all sound systems need to be complex with separate components connected together with cables. An all-in-one portable PA system, such as the one in *Figure 8-7*, is ideal for sound systems on the go — either for indoor or outdoor use. The portable PA system has its own amplifier and speaker, and includes controls to adjust the gain and tonal balance of the sound. Power output is modest — most portable PA systems are rated at 10 to 20 watts — which is sufficient for small gatherings. The input of the amplifier accepts either line-level, microphone, or electric guitar (magnetic) pickup.

Portable PA systems are also ideally suited as "monitor speakers" for use by the performer. A monitor speaker is placed on the stage, usually directly in front of and facing the performer. The speaker helps the performer hear what he or she is playing, and is especially important in rock music presentations when the sound level is generally very high. To use a portable PA system as a monitor speaker, connect an output from the mixing console to the input of the portable PA system (use a Y-adapter if your mixing console only has one output jack).

HANDS-FREE INTERCOM SYSTEM

A hands-free intercom system lets the technical support technicians communicate with one another during a performance. For example, as a sound system operator, you will likely need to coordinate your efforts with the lighting system operator. If you're involved in a theatrical production, the stage manager must be in constant contact with the sound and light system operators as well as backstage helpers.

A typical, relatively inexpensive, hands-free intercom system is based on walkie-talkies that use the FM broadcast band. Each unit, which is normally in the receive mode, has a headset with headphones and boom microphone. There is no "push to talk" button because they are voice-actuated; when you talk, it automatically switches to transmit mode, leaving your hands free to operate the controls of the sound system. This type of system allows you to expand the number of users simply by adding more units. The only catch: only one person can talk at a time.

Some other hands-free intercom systems operate in duplex mode; that is, they allow two — and sometimes three — people to talk at the same time (like the usual telephone system). These systems are always in transmit mode. This type of hands-free intercom system does not permit easy expansion to add more users.

HOME THEATER SOUND SYSTEM

For several years, home owners have thought that the picture from a TV is only half the story; the other half is the sound. From this, the "home theater" has become popular. More and more living rooms are being redesigned as miniature movie houses, with big-screen TVs, stereo amplifiers, and a matrix of carefully placed speakers.

PUBLIC THEATER SYSTEM

The home theater concept can be employed in corporate conference and training rooms, as well as in small auditoriums and meeting halls. Many corporate conference rooms are going high-tech and even include large-screen television sets and stereo systems. Entertainment isn't the main objective here — the boss doesn't want you watching TV programs or movies on the company's projection TV set. Rather, the audio/video system in the conference room provides a convenient teleconferencing center. The system can also be used for training or video demonstrations.

A public theater audio and video system, shown in *Figure 8-8,* consists of the following:

- **Large screen TV.** The screen must be large enough so that everyone in the room can easily see the picture. The larger the room, the larger the screen needed. In most cases, a screen measuring no less than 35 inches diagonally is required for public theater setup. Direct view screens are available up to about 40 inches. Larger screens are available with projection units; some use rear projection and some use front projection.

- **A/V receiver with Dolby® Pro-Logic® decoder.** The A/V receiver amplifies the sound from the source (VCR, video disc, satellite receiver, etc.). The Dolby Pro-Logic decoder directs the sound to the proper speakers. Most higher-end A/V receivers now have the Dolby Pro-Logic decoder built-in.

- **Two main front speakers.** These speakers flank the TV, providing stereo sound from the front of the room. Typically, the speakers are placed no more than 5-10 feet from either side of the TV or screen.

- **Center channel speaker.** Part of the Dolby Pro-Logic system is support for a center channel speaker, usually placed directly on top of the TV or screen. This speaker makes the sound from the TV more natural — the sound from the people on the screen actually seems to come from the screen. Center-channel

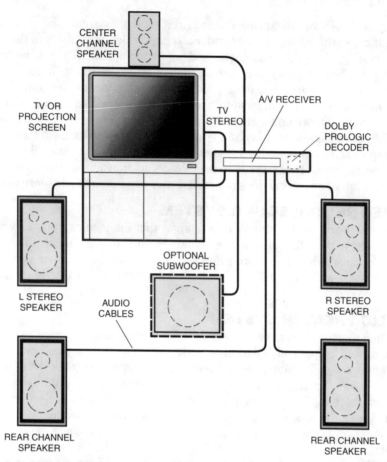

Figure 8-8. A typical public and home theater setup includes a large-screen TV or monitor, two stereo speakers, a center channel speaker, and a pair of smaller rear-channel speakers. An A/V receiver with Dolby Pro-Logic circuits provides the appropriate inputs for each unit.

speakers are most often equipped with magnetic shielding, so that the magnetic field from the heavy magnet used in the speaker does not interfere with the TV set (or worse, your VCR and its videotapes!). For best results, this speaker must be of high quality.

- **Rear channel speakers.** Additional speakers are located at the back of the room to add ambience. Though two speakers are used in the Dolby Pro-Logic system, both are fed the same monaural signal. Rear channel speakers are typically small and mounted on the wall.

More elaborate public (and home) theater systems also have a subwoofer speaker situated near the front of the room. The subwoofer provides very low-frequency sound. Only one subwoofer is required because our ears cannot detect the direction of very low-frequency sound signals. Most subwoofers are very large physically and the heavy vibrations of the low-frequency sound require very sturdy construction. Some subwoofers also serve as coffee tables!

ONE SYSTEM: THREE WAYS TO LISTEN TO IT

The Dolby Pro-Logic sound system supports several listening modes. These modes adapt the sound as needed.

- Dolby Pro-Logic mode is used when the program source — from a VCR, for example — is recorded in Dolby Surround. (Note that Dolby Surround is a completely different technique from Dolby Noise Reduction (DNR); the two have nothing to do with one another.) This mode provides the best over-all sound. The center channel speaker carries the bulk of the spoken dialog. Music and extra dialog come from the front left and right speakers. Sound effects and other ambience (including music) come from the rear channel speakers. Many A/V receivers provide an option to bypass the center channel speaker and direct the sound to the right and left speakers instead (this is often called "phantom mode").

- Stereo mode is used to listen to standard stereo sound. The center channel speaker and rear channel speakers are not used. This mode is used when the program is not recorded in Dolby Surround.

- Stereo with effects mode (also called "Hall," "Surround," or "3-way") adds artificial delay to enhance the fullness of the sound. The front stereo speakers carry the bulk of the sound; a small amount of digitally delayed sound is piped to the rear speakers. This causes a reverb which the listener perceives as a richer and fuller sound. Use this mode when the program is not recorded in Dolby Surround, but you wish to add ambience to the audio portion of the presentation.

Note that there are other methods of delivering sound to multiple speakers. The Dolby Pro-Logic system is the most common in home and public theater setups because of its relative low cost and flexibility. Other systems, which at this writing are primarily used only in movie theaters and very high-end home theater setups, are based on digital audio processes and more accurately steer signals to their destination speakers. For example, there's Dolby AC-3, DTS, and SDDS — none of these systems is compatible with the other. Programs can support more than one only if they carry additional digital sound tracks.

TIME TO ENJOY YOUR SOUND SYSTEM

By now you may have already purchased, installed, and operated your sound system. No doubt it will bring you many years of faithful service. As you use your sound system, keep in mind its basic purpose: to let everyone in the room hear the program — whether it's a musical act, a lecture, or a dramatic play — with the least amount of noise and distortion. Sound systems become counter-productive when their users forget this simple fact. If this book teaches you only one thing about effectively using a sound system it should be this: *It's the quality of sound that matters, not its loudness.*

Sound System Maintenance Journal

Vital Statistics:

	Brand & Model	Specifications	Serial Number
Microphone 1			
Microphone 2			
Microphone 3			
Microphone 4			
Microphone 5			
Microphone 6			
Loudspeakers, sub			
Loudspeakers, main			
Mixing Console			
Power amplifier			
Other			
Other			
Other			

PM Schedule Supplies:

Household spray cleaner _____

Non-petroleum solvent cleaner _____

Contact cleaner _____

Other: _____

_____

_____

General Maintenance:

Date _____

    Exterior clean
    Interior clean
    Clean/inspect connectors
    Inspect cables
    Inspect microphone battery

Repair/Replacement Record:

Date _____

Item repaired/replaced _____

Reason for repair/replacement _____

_____

Action taken _____

_____

Notes _____

_____

_____

Repair/Replacement Record:

Date _____

Item repaired/replaced _____

Reason for repair/replacement _____

_____

Action taken _____

_____

Notes _____

_____

_____

Repair/Replacement Record:

Date _____

Item repaired/replaced _____

Reason for repair/replacement _____

_____

Action taken _____

_____

Notes _____

_____

_____

A. Logarithms

EXPONENTS

A logarithm (log) is the exponent (or power) to which a given number, called the base, must be raised to equal the quantity. For example:

Since $10^2 = 100$, then the log of 100 to the base 10 is equal to 2, or $\text{Log}_{10} 100 = 2$

Since $10^3 = 1000$, then the log of 1000 to the base 10 is equal to 3, or $\text{Log}_{10} 1000 = 3$

BASES

There are three popular bases in use—10, 2 and ϵ. Logarithms to the base 10 are called common logarithms (log). Logarithms in base ϵ are called natural logarithms (ln).

Logarithms to the base 2 are used extensively in digital electronics.

Logarithms to the base ϵ (approximately 2.71828...) are quite frequently used in mathematics, science and technology. Here are examples:

Base 10

$\log_{10} 2 = 0.301$ is $10^{0.301} = 2$

$\log_{10} 200 = 2.301$ is $10^{2.301} = 200$

Base 2

$\log_2 8 = 3$ is $2^3 = 8$

$\log_2 256 = 8$ is $2^8 = 256$

Base ϵ

$\text{in}_\epsilon 2.71828 = 1$ is $\epsilon^1 = 2.71828$

$\text{in}_\epsilon 7.38905 = 2$ is $\epsilon^2 = 7.38905$

RULES OF EXPONENTS

Since a logarithm is an exponent, the rules of exponents apply to logarithms:

$\log (M \times N) = (\log M) + (\log N)$

$\log (M/N) = (\log M) - (\log N)$

$\log M^N = N \log M$

B. Decibels

The bel is a logarithmic unit used to indicate a ratio of two power levels (sound, noise or signal voltage, microwaves). It is named in honor of Alexander Graham Bell (1847-1922) whose research accomplishments in sound were monumental. A 1 bel change in strength represents a change of ten times the power ratio. In normal practice, the bel is a rather large unit, so the decibel (dB), which is 1/10 of a bel, is commonly used.

Number of dB = 10 log P2/P1

A 1 dB increase is an increase of 1.258 times the power ratio, or 1 db = 10 log 1.258.

A 10 dB increase is an increase of 10 times the power ratio, or 10 db = 10 log 10.

Other examples are:

3 dB = 2 times the power ratio

20 dB = 100 times the power ratio

-30 dB = 0.001 times the power ratio

It is essential to remember that the decibel is *not* an absolute quantity. It merely represents a change in power level relative to the level at some different time or place. It is meaningless to say that a given amplifier has an output of so many dB unless that output is referred to a specific power level. If we know the value of the input power, then the *ratio* of the output power to the specific input power (called power gain) may be expressed in dB.

If a standard reference level is used, then *absolute power* may be expressed in dB *relative* to that standard reference. The commonly used reference level is one milliwatt. Power referenced to this level is expressed in dBm. Here are power ratios and dBm ratios:

dB	Power Ratio	dBm	Power (mw)
1	1.258	1	1.258
3	2	3	2
10	10	10	10
20	100	20	100
−30	0.001	−30	0.001

Glossary

absorption: The effect of a material that reflects only a portion of the sound waves striking it. Common absorptive materials include carpeting, acoustic tile, and drapes.

acoustic feedback: A squealing sound when the output of an audio circuit is fed back in phase into the circuit's input.

ac coupling: Coupling between electronic circuits that passes only alternating current and time varying signals, not direct current.

acoustics: The science or study of sound.

alternating current (ac): An electrical current that periodically changes in magnitude and direction.

ambience: A surround or concert-hall sound.

ampere (A): The unit of measurement for electrical current in coulombs (6.25 x 10^{18} electrons) per second. There is one ampere in a circuit that has one ohm resistance when one volt is applied to the circuit. See Ohm's Law.

amplifier: An electrical circuit designed to increase the current, voltage, or power of an applied signal. As used in sound systems, an amplifier (also a power amplifier) increases the power of an electrical signal from a microphone or other source, and applies the amplified signal to one or more speakers, so that it can be heard by a group of people.

amplitude: The relative strength (usually voltage) of a signal. Amplitude can be expressed as either a negative or positive number, depending on the signals being compared.

attenuation: The reduction, typically by some controlled amount, of an electrical signal.

audio frequency: The acoustic spectrum of human hearing, generally regard to be between 20 Hz and 20,000 Hz.

baffle: A piece of wood inside or outside a speaker enclosure to direct or block the movement of sound.

balance: Equal signal strength provided to both left and right stereo channels.

balanced line: A circuit using two identical conductors operated so the voltages on each of them are equal in magnitude but opposite in polarity with respect to ground. See also unbalanced line.

bandpass filter: An electric circuit designed to pass only middle frequencies. See also high-pass filter and low-pass filter.

bass: The low end of the audio frequency spectrum: approximately 20 Hz to about 1000 Hz.

capacitor (C): A device made up of two metallic plates separated by a dielectric (insulating material). Used to store electrical energy in the electrostatic field between the plates. It produces a capacitive reactance which is an impedance to an ac current.

cassette: The two-reel plastic carrier that contains audio magnetic tape, for recording or playback of audio information.

CD: Compact disc, or compact disc player.

channel: The left or right signals of a stereo audio system.

circuit: A complete path that allows electrical current from one terminal of a voltage source to the other terminal.

clipping: A distortion caused by cutting off the peaks of audio signals. Clipping usually occurs in the amplifier when its input signal is too high or when the volume (gain) control is turned up too high.

coloration: Basically a change in the frequencies of an original sound signal. In audio, the effect is "smearing" sounds by adding frequencies due to intermodulation distortion. It is most prevalent at high audio frequencies.

compliance: The relative stiffness of a speaker suspension, typically indicated simple as "high" or "low," but technically specified as V_{as}.

crossover network: An electric circuit or network that splits the audio frequencies into different bands for application to individual speakers.

current (I): The flow of charge measured in amperes.

decibel (dB): A logarithmic scale used to denote a change in the relative strength of an electric signal or acoustic wave. It is a standard unit for expressing the ratio between power level P_1 and power level P_2 {dB = 10 $\log_{10} P_1/P_2$}. An increase of 3 dB is a doubling of electrical (or signal) power; an increase of 10 dB is a doubling of perceived loudness. The decibel is not an absolute measurement, actually, but indicates the relationship or ratio between two signal levels.

direct current (dc): Current in only one direction.

dispersion: The spreading of sound waves as they leave a speaker.

distortion: Any undesirable change in the characteristics of an audio signal.

Dolby: Trademarked name for a variety of sound processing circuitry, including noise reduction (Dolby Noise Reduction) and surround sound (Dolby Pro-Logic).

driver: The electromagnetic components of a speaker, typically consisting of a magnet and voice coil.

dynamic range: The range of sound levels which a system can reproduce without distortion. When expressed in decibels, the ratio between the softest and loudest sound levels in a signal range or that an amplifier can reproduce.

echo: A reflected sound wave of sufficient amplitude and delay to make it distinct from the original sound.

equalizer: An adjustable audio filter inserted in a circuit to divide and adjust its frequency response.

equalization: As used in audio, the adjustment of frequency response to tailor the sound to match personal preferences, room acoustics, and speaker enclosure design.

farad: The basic unit of capacitance. A capacitor has a value of one farad when it can store one coulomb of charge with one volt across it. One farad is a considerable amount of storage; most capacitors are rated in millionths and even billionths of a farad.

feedback: see acoustic feedback.

fidelity: A measure of how true a circuit, amplifier, system, or subsystem reproduces its input signal.

filter: An electrical circuit designed to prevent or reduce the passage of certain frequencies.

flat response: The faithful reproduction of an audio signal; specifically, variations in output level of less than one decibel above or below a median level over the audio spectrum.

frequency: The number of waves (or cycles) arriving at or passing a point in one second; expressed in hertz (or Hz).

frequency distortion: The distortion produced when certain frequencies are amplified more or less than others to produce frequencies in the output that are not present in the input.

frequency response: The range of frequencies that are faithfully reproduced by a given sound system or speaker.

full-range: A speaker designed to reproduce all or most of the sound spectrum.

gain (G): The increase in magnitude of an electrical signal, usually expressed as an output divided by an input.

ground: Refers to a point of (usually) zero voltage, and can pertain to a power circuit or a signal circuit.

harmonic: The multiple frequencies of a given sound, created by the interaction of signal waveforms. A "middle C on the piano has a fundamental audio frequency of 256 Hz, but also a number of secondary higher frequencies (harmonics) that are odd and even multiples of this fundamental.

harmonic distortion: Harmonics artificially added by an electrical circuit or speaker, and are generally undesirable. It is expressed as a percentage of the original signal.

headroom: The amount of dynamic range between the average sound pressure level and the absolute maximum sound level.

hertz: A unit of frequency equal to one cycle per second, named after German physicist H.R. Hertz.

high-pass filter: An electric circuit designed to pass only high frequencies. See also bandpass filter and low-pass filter.

hiss: Audio noise that sounds like air escaping from a tire, typically caused by thermal noise generated by an electric circuit.

hum: Audio noise that has a steady low frequency pitch, typically caused by the effects of induction by nearby ac lines or leakage of ac line frequency into an amplifier's signal circuits.

impedance (Z): The opposition of a circuit or speaker to an alternating current.

inductance (L): The capability of a coil to store energy in a magnetic field surrounding it. It produces an inductive reactance which is an impedance to an ac current.

infrasonic: Sound frequencies below the 20-Hz threshold of normal human hearing; formerly called subsonic. See also ultrasonic.

intensity: The strength of a sound signal represented by the amplitude of the pressure waves producing the sound. Expressed in watts per square meter.

intermodulation distortion (IM): The distortion that occurs when an electric circuit develops new frequencies (sums and differences of original signals) from the ones being processed.

inverse square law: A law which states that the level of sound decreases 50 percent for every doubling of the distance from the sound source.

loudness: A measure of the sensitivity of human hearing to the strength of sound. Often used as a relative measure of the intensity of sound.

low-pass filter: An electric circuit designed to pass only low frequencies. See also bandpass filter and high-pass filter.

microphone: A device that converts sound waves into electrical signals.

midrange: A speaker designed to reproduce the middle frequencies of the sound spectrum, generally most efficient between about 1000 Hz to 4000 Hz.

mixing console: An electronic device that allows for the combination and manipulation of multiple signal inputs, such as microphones and tape decks. The output of the mixing console is connected to an amplifier.

noise: An unwanted sound.

noise floor: The ambient sound and system noise below the program sound

ohm (Ω): A unit of electrical resistance or impedance.

Ohm's law: A basic law of electric circuits. It states that the current I in amperes in a circuit is equal to the voltage E in volts, divided by the resistance R in ohms; thus, $I = E/R$.

pan: A control on a mixing console or power amplifier that directs the signal to left and right stereo channels in varying degrees. Also called balance.

peak: The maximum amplitude of a voltage or current.

period: For electronic circuits, the length of time required for one cycle of a periodic wave.

phase: The angular or time displacement between the voltage and current in an ac circuit.

phase distortion: The distortion which occurs when one frequency component of a complex input signal takes longer to pass through an amplifier or system than another frequency.

piezoelectric: A characteristic of some materials, especially crystal, that when subjected to electric voltage, the material vibrates. Often used in certain speaker designs, such as high-frequency tweeters. Can also be used in reverse: sound vibrations generate an electric current. In this way, the piezoelectric material is used as a microphone.

polarity (electrical circuits): In electrical circuits, the description of whether a voltage is positive or negative with respect to some reference point.

polarity (radiated fields): The orientation of magnetic or electric fields. The polarity of the incoming audio signal determines the direction of movement of a speaker cone.

power: The time rate of doing work or the rate at which energy is used. A watt of electrical power is the use of one joule of energy per second. Watts of electrical power equals volts times amperes.

public address (PA) system: A sound system specifically designed to amplify sound signals (usually speech) throughout a large, often non-contiguous, area.

resonance: The tendency of a speaker to vibrate most at a particular frequency; sometimes referred to as natural frequency.

resistance: In electric or electronic circuit, a characteristic of a material that opposes the flow of electrons. It results in loss of energy in a circuit dissipated as heat.

RMS: An acronym for root mean square. The RMS value of an alternating current produces the same heating effect in a circuit as the same value of a direct current.

reverberation: Echo-like repetition of an original sound signal. The reverberation can be caused by physical boundaries, such as sound bouncing off a wall. Reverberation can also be created electronically, and in this case is often used to add "depth" to a sound signal.

sensitivity: In sound systems, a measure of the ability of a transducer (such as a microphone) to adequately detect a signal so that it can be amplified. In a receiver or tuner, the ability to amplify very weak signals.

signal: The desired portion of electrical information.

signal-to-noise (S/N): The ratio, expressed in dB, between the signal (sound you want) and noise (sound you don't want).

sine wave: The waveform of a pure alternating current or voltage. It deviates about a zero point to a positive value and a negative value. Audio signals are sine waves or combinations of sine waves.

sound pressure level (SPL): The loudness of an acoustic wave stated in DB that is proportional to the logarithm of its intensity.

sound reinforcement: The generic term for sound systems used for public address, public performances, speeches, etc.

sound spectrum: The range of sound frequencies discernable by the human ear, generally accepted as 20 Hz to 20,000 Hz. In actuality, sound waves can exist with frequencies just above zero Hz (infrasonic or subsonic), and well above the range of human hearing (ultrasonic).

static: Random noise in a sound system due to atmospheric or manmade electrical disturbances, such as lightning.

subsonic: See infrasonic.

subwoofer: A speaker specifically designed to reproduce extremely low frequency signals, from about 100 Hz to near zero Hz. Because of the very low frequencies involved, subwoofers often create a "sensation" of sound and air movement, rather than a audible tone.

surround: As used in home/public theater sound systems, a generic term for the speakers placed to the rear of the listening audience, to provide sound ambience. As used in speakers, the outer suspension of a speaker cone.

three-way (3-way): A type of speaker systems composed of three ranges of speakers,

specifically a tweeter, midrange, and woofer. See also two-way.

timbre: A subjective term used for human hearing that gives a sound a particular identity. It is related to the spectrum of frequencies contained within a sound.

total harmonic distortion (THD): The percentage, in relation to a pure input signal, of harmonically derived frequencies introduced in the sound reproducing circuitry (including the speakers).

treble: The upper end of the audio spectrum, usually reproduced by a tweeter.

transducer: A device that converts an original sound signal (like a voice) into electrical energy so that it can be amplified. Typical transducers include microphones and magnetic guitar pickups.

transient intermodulation distortion: A distortion which occurs principally during loud, high-frequency music passages in solid-state amplifiers that use large amounts of negative feedback.

transient response: The instantaneous change in an electronic circuit's output response when input circuit conditions suddenly change from one steady-state condition to another.

tweeter: A speaker designed to reproduce the high or treble range of the sound spectrum, generally most efficient from about 4000 Hz to 20,000 Hz.

two-way (2-way): A type of speaker systems composed of two ranges of speakers, usually tweeter and midrange, or midrange and woofer. See also three-way.

ultrasonic: Sound frequencies above the 20,000 Hz threshold of normal human hearing. See also subsonic.

unbalanced line: Any transmission line in which the two conductors are at different potential with respect to ground. See also balanced line.

wavelength: The distance a wave travels in the time required to complete one cycle.

watt: A unit of electrical power.

woofer: A speaker designed to reproduce the low frequencies of the sound spectrum, generally most efficient from about 20 Hz to 1000 Hz.

Index

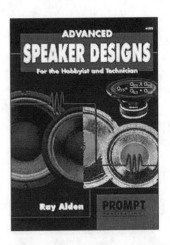

PROMPT Publications is your best source for informative books in the technical field.

Sound Systems for Your Automobile

The How-To Guide for Audio System Selection and Installation
by Alvis J. Evans and Eric J. Evans
Whether you're starting from scratch or upgrading, this book will show you how to plan your car stereo system, choose components and speakers, and install and interconnect them to achieve the best sound quality possible. Easy-to-follow steps, parts lists, wiring diagrams, and fully illustrated examples make planning and installing a new system easy.
$16.95
Paper/124 pp./6 x 9"/Illustrated
ISBN#: 0-7906-1046-9
Pub. Date 1/94

Security Systems for Your Home and Automobile

by Gordon McComb
Because of the escalating threat of theft and violence in today's world, planning, selecting, and installing security systems to protect your home and automobile is vital. You can save money by installing a system yourself. In simple, easy-to-understand language, *Security Systems for Your Home and Automobile* tells you everything you need to know to select and install a security system with a minimum of tools.
$16.95
Paper/130 pp./6 x 9"/Illustrated
ISBN#: 0-7906-1054-X
Pub. Date 7/94

Call 1-800-428-7267 TODAY for the name of your nearest PROMPT Publications distributor. Be sure to ask for your FREE PROMPT catalog!

PROMPT Publications is your best source for informative books in the technical field.

VOM and DVM Multitesters for the Hobbyist and Technician

by Alvis J. Evans
VOM and DVM Multitesters offers concise, clearly illustrated text to explain how digital and analog meters work, as well as their uses on the job, in the workshop, and in the home. Subjects include basic concepts of VOM and DVM meters, multitester measurements, and meter troubleshooting.

$14.95
Paper/144 pp./6 x 9"/Illustrated
ISBN#: 0-7906-1031-0
Pub. Date 9/92

The Multitester Guide
How to Use Your Multitester for Electrical Testing and Troubleshooting
by Alvis J. Evans
With the instructions provided in *The Multitester Guide*, basic electrical measurements become easily, correctly, and quickly completed. In addition to the functions and uses of multitesters, the easy-to-understand text and clear examples cover such topics as the measurement of basic electric components and their in-circuit performance; the measurement of home lighting, appliance, and related systems; and automotive circuit measurements.

$14.95
Paper/160 pp./6 x 9"/Illustrated
ISBN#: 0-7906-1027-2
Pub. Date 12/92

Call 1-800-428-7267 TODAY for the name of your nearest PROMPT Publications distributor. Be sure to ask for your FREE PROMPT catalog!

PROMPT Publications is your best source for informative books in the technical field.

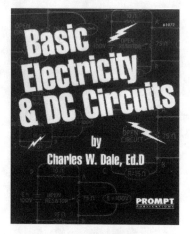

Tube Substitution Handbook

Complete Guide to Replacements for Vacuum Tubes and Picture Tubes
by William Smith & Barry Buchanan
The most accurate, up-to-date guide available, the *Tube Substitution Handbook* is useful to antique radio buffs, old car enthusiasts, ham operators, and collectors of vintage ham radio equipment. In addition, marine operators, microwave repair technicians, and TV and radio technicians will find the *Handbook* to be an invaluable reference tool. Diagrams are included as a handy reference to pin numbers for the tubes listed in the *Handbook*.

$16.95
Paper/ 149 pp./6 x 9"/Illustrated
ISBN#: 0-7906-1036-1
Pub. Date 12/92

Basic Electricity & DC Circuits

by Charles W. Dale, Ed.D
Electricity is constantly at work around your home and community, lighting rooms, running manufacturing facilities, cooling stores and offices, playing radios and stereos, and computing bank accounts. Now you can learn the basic concepts and fundamentals behind electricity and how it is used and controlled. *Basic Electricity and DC Circuits* shows you how to predict and control the behavior of complex DC circuits. Concepts and terms are introduced as you need them, with many detailed examples and illustrations.

$34.95
Paper/928 pp./6 x 9"/Illustrated
ISBN#: 0-7906-1072-8
Pub. Date 8/95

Call 1-800-428-7267 TODAY for the name of your nearest PROMPT Publications distributor. Be sure to ask for your FREE PROMPT catalog!

PROMPT Publications is your best source for informative books in the technical field.

The In-Home VCR Mechanical Repair & Cleaning Guide
by Curt Reeder

You can save money and learn valuable skills repairing your VCR in your own home! *The In-Home VCR Mechanical Repair & Cleaning Guide* is geared toward the average home VCR user who owns one or more VCRs and would like assistance in VCR maintenance. This book presents repairing & cleaning VCRs as a step-by-step process. This book belongs in the library of anyone who ever wanted to know how to repair their own VCR or understand its workings.

$19.95
Paper/240 pp./8-1/2 x 11"/Illustrated
ISBN#: 0-7906-1076-0
Pub. Date 3/96

Schematic Diagrams
The Basics of Interpretation and Use
by J. Richard Johnson

Step-by-step, *Schematic Diagrams* shows you how to recognize schematic symbols and their uses and functions in diagrams. You will also learn how to interpret diagrams so you can design, maintain, and repair electronics equipment. Subjects covered include component symbols and diagram formation, functional sequence and block diagrams, power supplies, audio system diagrams, computer diagrams, and more.

$16.95
Paper/208 pp./6 x 9"/Illustrated
ISBN#: 0-7906-1059-0
Pub. Date 9/94

Call 1-800-428-7267 TODAY for the name of your nearest PROMPT Publications distributor. Be sure to ask for your FREE PROMPT catalog!

PROMPT Publications is your best source for informative books in the technical field.

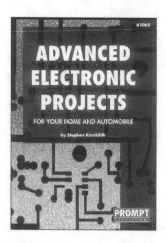

Advanced Electronic Projects for Your Home and Automobile

by Stephen Kamichik
You will gain valuable experience in the field of advanced electronics by learning how to build the interesting and useful projects featured in *Advanced Electronic Projects*. The projects in this book can be accomplished whether you are an experienced electronic hobbyist or an electronic engineer, and are certain to bring years of enjoyment and reliable service.

$18.95
Paper/160 pp./6 x 9"/Illustrated
ISBN#: 0-7906-1065-5
Pub. Date 5/95

Electronic Control Projects for the Hobbyist and Technician

by Henry C. Smith & Craig B. Foster
Would you like to know how and why an electronic circuit works, and then apply that knowledge to building practical and dependable projects that solve real, everyday problems? Each project in *Electronic Control Projects* involves the reader in the actual synthesis of a circuit. A complete schematic is provided for each circuit, along with a detailed description of how it works, component functions, and troubleshooting guidelines.

$16.95
Paper/168 pp./6 x 9"/Illustrated
ISBN#: 0-7906-1044-2
Pub. Date 11/93

Call 1-800-428-7267 TODAY for the name of your nearest PROMPT Publications distributor. Be sure to ask for your FREE PROMPT catalog!

☞ **Dear Reader:** *We'd like your views on the books we publish.*

PROMPT® Publications, an imprint of Howard W. Sams & Company, is dedicated to bringing you timely and authoritative documentation and information you can use. You can help us in our continuing effort to meet you information needs. Please take a few moments to answer the questions below. Your answers will help us serve you better in the future.

1. What is the title of the book you purchased?_____

2. Where do you usually buy books? _____

3. Where did you buy this book? _____

4. What did you like most about the book? _____

5. What did you like least? _____

6. Is there any other information you'd like included? _____

7. In what subject areas would you like us to publish more books? (Please check the boxes next to your fields of interest.)

❏ Audio Equipment Repair   ❏ Home Appliance Repair

❏ Camcorder Repair   ❏ Mobile Communications

❏ Computer Hardware   ❏ Security Systems

❏ Electronic Concepts Theory   ❏ Sound System Installation

❏ Electronic Projects/Hobbies   ❏ TV Repair

❏ Electronic Reference   ❏ VCR Repair

8. Are there other subjects that you'd like to see books about? _____
_____

9. Comments _____
_____
_____

Name _____
Address _____
City _____ State/ZIP _____
Online Address _____
Would you like a *FREE* PROMPT® Publications catalog? ❏Yes ❏No
Thank you for helping us make our books better for all of our readers. Please drop this postage-paid card into the nearest mailbox.

For more information about PROMPT® Publications,
see your authorized Howard Sams distributor, or call 1-800-428-7267
for the name of your nearest PROMPT® Publications distributor.

An imprint of
Howard W. Sams & Company
A Bell Atlantic Company
2647 Waterfront Parkway, East Dr.
Suite 300
Indianapolis, IN 46214-2041